SOUND SYSTEMS FOR YOUR AUTOMOBILE

By

Alvis J. Evans, Eric J. Evans

Contributing Technical Editor

Joseph A. D'Appolito, Ph.D
Audio and Loudspeaker Design Consultant

PROMPT
PUBLICATIONS
An Imprint of
Howard W. Sams & Company
Indianapolis, Indiana

REVISED FIRST EDITION, 1993

PROMPT® Publications is an imprint of Howard W. Sams & Company, 2647 Waterfront Parkway, East Drive, Indianapolis, IN 46214-2041.
This book was originally developed and published as *Automotive Sound Systems* *by:*

Master Publishing, Inc.
14 Canyon Creek Village MS31
Richardson, TX 75080
(214) 907-8938

International Standard Book Number: 0-7906-1046-9

Edited by: *Charles Battle, Gerald Luecke*
Cover Design by: *Sara Wright*
Contributions by: *Jim Imboden, Consumer Products Division, Ovation Audio/Video Specialists, Radio Shack*
Design and artwork by: *Plunk Design, Dallas, TX*

Acknowledgements:
Eric Evans expresses appreciation to Kirk Conger of Sound Works, Fort Worth, TX for his assistance, and Angela Evans for her review.
All photographs not credited are courtesy of Author, Ovation Audio/Video Specialists, Radio Shack, Master Publishing, Inc., or Howard W. Sams & Company.

Printed in the United States of America
9 8 7 6 5 4 3 2

TABLE OF CONTENTS

PREFACE

Drivers of automotive vehicles are looking for automotive sound systems with increased performance. They want to enjoy the best sound that the radio, cassette, and CD media have to offer. That want is being supported by a rapidly expanding audio technology industry that is providing sophisticated equipment at affordable prices. Yet vehicle manufacturers, except in top-of-the-line models or through special high-priced options, have not kept pace with the latest available technology.

Vehicle owners do have a solution. They can use the same amount of money charged by the vehicle manufacturer for the factory-installed sound system and buy an aftermarket system that has more power, features, and notably better quality sound. Or they can enhance their present system by substituting up-graded components, or adding on components. There is no reason why an automotive sound system should not provide the occupants with "live performance" reproductions that rival home audio systems.

This may seem like a formidable task to the average vehicle owner. Sound system components can be expensive, and the systems seem complicated. *Sound Systems for Your Automobile* shows how to do it. It was written for that purpose. It discusses actual installations—in coupes/sedans, hatchbacks, pickups, sport utility vehicles and vans—from simple add-ons to completely replaced systems. *Sound Systems for Your Automobile* is a proponent of planned systems that can be expanded step-by-step as finances become available.

Chapter 1 begins with terms and definitions that establish an understanding of sound fundamentals. If you are familiar with this material, you may bypass it.

Chapters 2, 3, and 4 discuss the individual components, their features, and the steps to follow for system improvement—from simple up-grading to replacing everything from scratch. Since speakers are such key components, Chapter 4 is devoted entirely to them.

Five chapters (5, 6, 7, 8, and 9) describe specific installations. Each sound system is planned to meet specific performance objectives in a specific vehicle. A parts list is provided, as well as a detailed discussion of the installation. For installations that require an enclosure, a detailed cutting layout is provided for the enclosure material.

Throughout the book, you should gain an understanding of the aftermarket components available, and the techniques for selecting and installing a high-quality system. Even though all vehicle makes and models are not covered, the installations chosen should provide sufficient information to adapt to any vehicle. The end result is a notably improved quality sound system. That was our goal; we hope we have succeeded.

AJE, EJE

SOUND FUNDAMENTALS 1

We live in a fast changing and unpredictable world. Almost every time we turn around, there is an innovative and better electronic product with features that we just cannot live without. For example, the automobile audio system has undergone many significant changes recently. If you like to hear good music while in your car, then you probably want a top quality automotive sound system. Today you can have just that. By following the suggestions and instructions in this book, you can install a system from scratch, or dramatically improve your existing system. Let's begin by looking at the terms that we will be using.

WHAT IS SOUND?

Sound can be defined in more than one way.[1] For our purposes, we can think of it as a wave of air with pressure alternations above and below a steady-state pressure. The source of the variations in air pressure is the physical movement of objects and surfaces in the air. Sound waves radiate out from the source in the same manner as a rock dropped in a pond of water causes waves to radiate from the rock's point of entry. For an electronic sound system, the speakers produce the alternating pressure variations.

TERMS AND DEFINITIONS

Frequency

Frequency is the rate at which the air pressure varies up and down. Slow variations, or low frequencies, produce the bass sounds. Fast vibrations, or high frequencies, produce the treble sounds and the crisp, sharp characteristic of music. Frequency is expressed in hertz, which means cycles per second.

The simplest form of periodic motion is the sine wave. *Figure 1-1* illustrates sine waves of different frequencies. The unique characteristic of a sine wave is that it exists as a single frequency, called a fundamental, without any harmonics or overtones (more about harmonics a little later). Notice that the sine wave is symmetrical, which means that the positive half of the waveform (the portion above the horizontal axis in the figure) is exactly the same as the negative half.

Intensity

The amplitude of the pressure variations, as shown in *Figure 1-1,* determines the intensity of sound. (Many people use the word "volume" to signify either the intensity of a sound or the magnitude of an audio frequency signal; however, "volume" is a layman's term and has no real technical meaning.)

[1] *Making Sense of Sound*, A.J. Evans, ©Copyright 1992, Master Publishing, Inc.

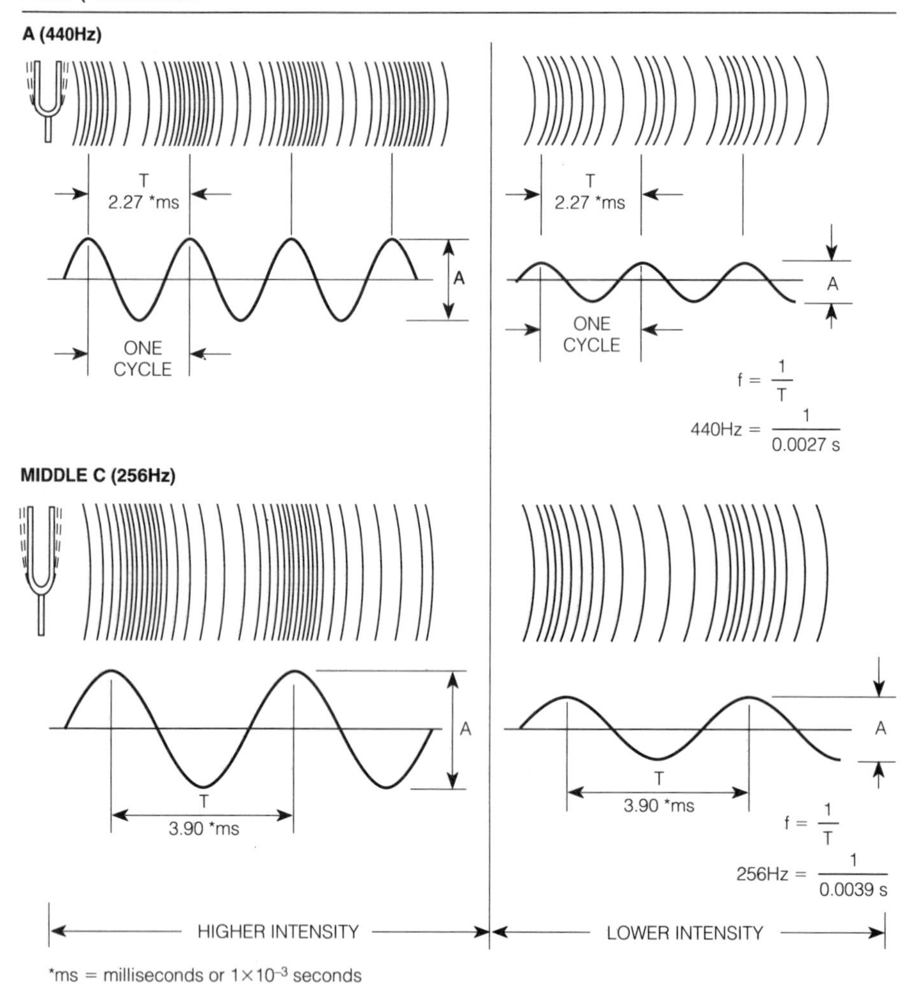

a. Pitch or Frequency is the Number of Vibrations Per Second. Intensity or Amplitude is the Amount of Energy Contained in the Wave.

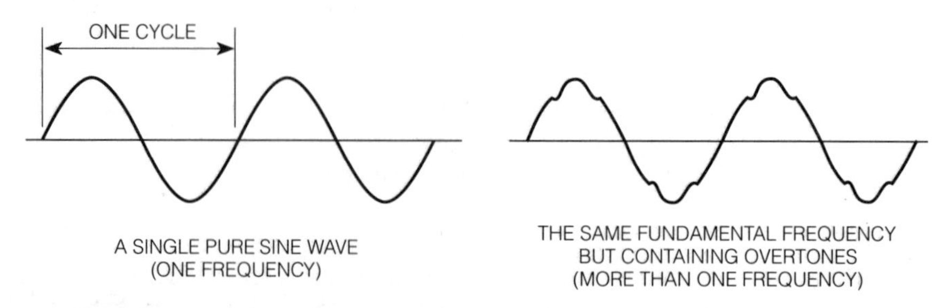

b. Timbre or Quality is the Purity of the Wave.

Figure 1-1. The motion of a tuning fork producing a wave in air and the corresponding sine wave associated with it.

Sound intensity is usually expressed in "decibels above the threshold of hearing" because loudness is approximately proportional to the logarithm of intensity (see discussion of decibels below). Sound intensity is most conveniently expressed in watts/square meter because the electrical power of the audio signal is in watts and the sound intensity depends on the total area over which this power is distributed. When comparing specifications of components, be sure that all are in the same units. Sometimes sound intensities are given in square centimeters and other times in square meters.

The human ear is extremely sensitive to low intensity sounds. It can detect a sound intensity as low as a tenth of a trillionth of a watt per square meter (10^{-13} W/m^2). For example, the total sound power generated by a large orchestra ranges from a few microwatts at the softest passages to many watts at the loudest passages.

Decibels

Acoustic intensity is commonly expressed in decibels (dB). Decibels are based on logarithms; however, we can present a few useful facts about the decibel without going into a mathematical derivation. The decibel is not an absolute measure of intensity. Rather it indicates a *ratio* between two sound intensity levels. The zero decibel (0dB) reference for sound intensities is one trillionth of a watt per square meter (10^{-12} W/m^2) at 1000Hz. The 0dB reference level represents the intensity of a pure 1000Hz tone that is just discernible by a person with normal hearing in an otherwise quiet background. This intensity level is also referred to as *threshold of hearing* at 1000Hz. (More on the threshold of hearing in a moment.)

For a power increase of ten times, the sound level increases 10dB. For a power increase of two times, the sound level increases 3dB. Conversely, when the power is halved, the sound level decreases 3dB. Because decibels are based on logarithms, when power multiplies, decibels add.

The threshold of hearing for humans changes with frequency. We are most sensitive to sounds in the 3 to 4kHz range where the threshold of hearing is actually negative 10dB. In other words, the intensity level that can just be heard between 3000 and 4000Hz is one tenth that which can just be heard at 1000Hz. We are least sensitive to low frequencies. The threshold of hearing is around 70dB at 30Hz. This means that the intensity level of a 30Hz tone must be ten million times greater than that of 1000Hz tone to be heard. This, at least in part, explains why so much more amplifier power is needed at low-frequencies for the sound to be heard.

CHARACTERISTICS OF SOUND

Sounds differ from one another in several basic physical properties, each of which has acoustic characteristics. The human ear can distinguish between two or more sounds when the sounds differ in one or more of the characteristics of loudness, pitch, or quality.

Loudness

The amount of energy in a wave determines its amplitude, which in turn, determines the maximum displacement of the vibrating particles of the medium. Loudness relates to the amplitude (A), or intensity, at which the energy is transmitted to the ear. We have already discussed intensity, so now let's look at pitch and quality.

Pitch

Pitch is primarily associated with frequency (f), which, as we discussed above, is the number of vibrations per second. The period of the wave is the time (T) in seconds required for a single cycle of the wave to pass a given point. As shown in *Figure 1-1,* f is the reciprocal of T and vice versa:

$$f = \frac{1}{T} \text{ and } T = \frac{1}{f}$$

Quality

Quality, or timbre, refers to the complexity of the wave. It is the main characteristic that distinguishes the sound of the same note (pitch) when played on two different instruments. The difference in the two sounds depends on the number of harmonics present and their amplitude relative to the amplitude of the fundamental. More on harmonics later, but, for now, harmonics are multiples of the fundamental frequency. The waveforms and the relative amplitude of the frequency components in three kinds of musical instruments — tuning fork, piano and clarinet — when playing a C note are illustrated in *Figure 1-2.*

Limits of Hearing

Not all sound waves can excite the sensation of hearing. Infrasonic waves are so low in frequency that the ear cannot hear them, but the body can feel the pressure variations. On the other end of the audio frequency spectrum are ultrasonic waves.

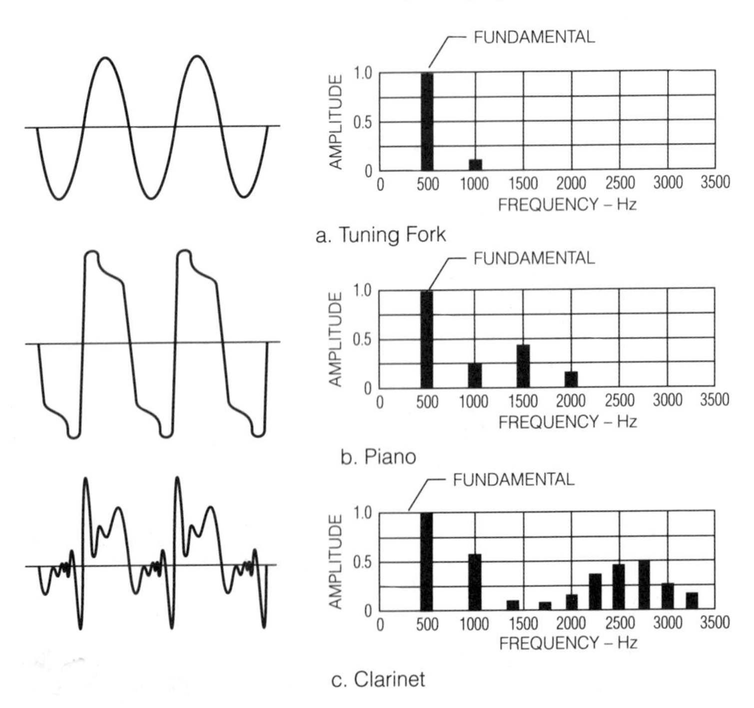

Figure 1-2. Waveforms and frequency distributions of tones produced by different musical instruments.

These frequencies are so high that they do not produce any sensation of sound in the human ear; however, a dog or an ultrasonic microphone can detect these frequencies. Some high-fidelity enthusiasts insist that their sound system be capable of reproducing frequencies above the human hearing range to maintain the high-frequency harmonic content and keep the timbre pure. The audio spectrum of various musical instruments is shown in *Figure 1-3.*

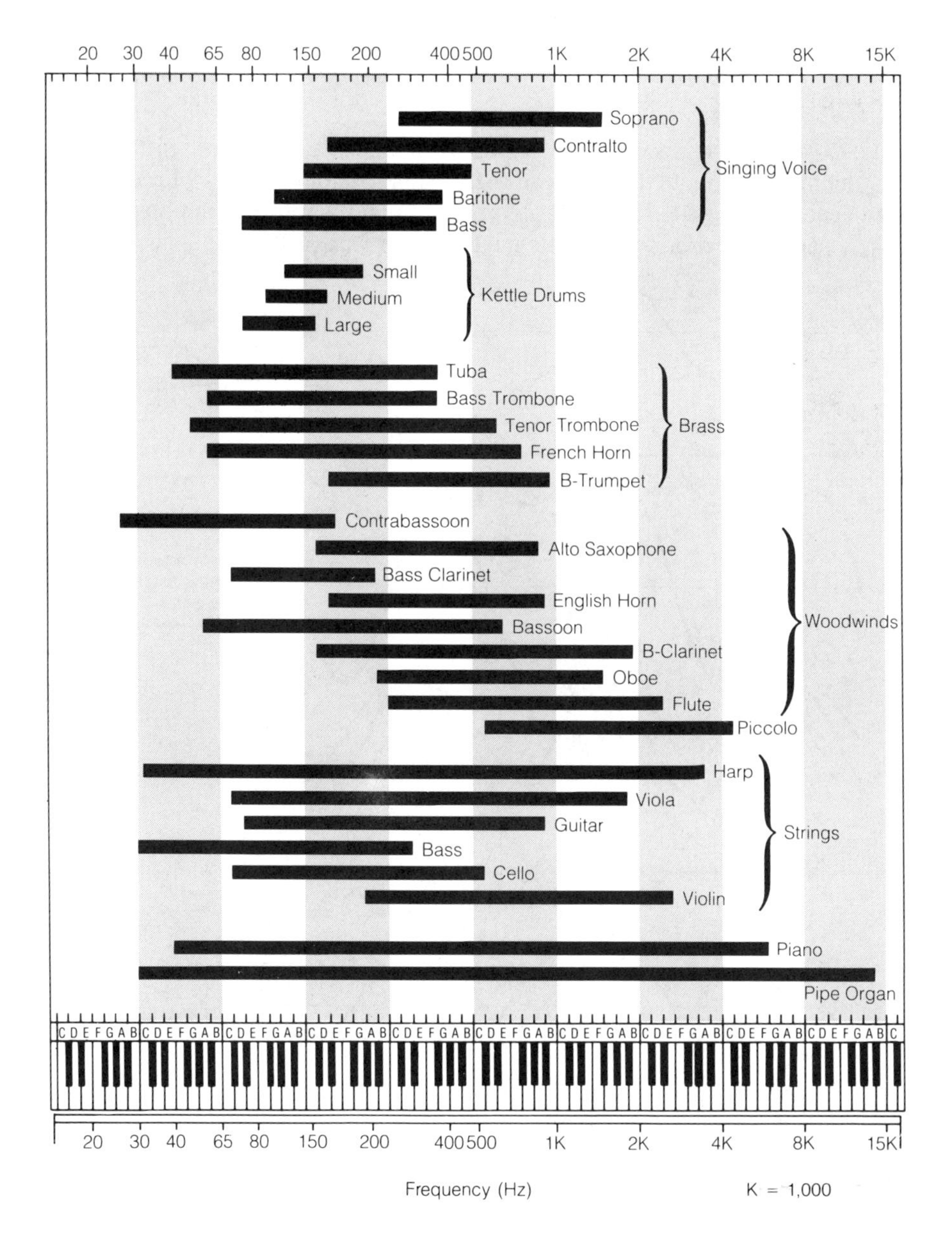

Figure 1-3. The audio spectrum showing the ranges of various instruments.

Phase

Two simple waves with the same frequency and amplitude may be different in phase. For a given frequency, phase is proportional to the time delay of one wave with respect to another. It can refer to either a sound wave or an electrical wave. Phase shift is often measured in degrees. If the two waves were combined after a phase shift, a new wave would result from constructive and destructive interference. *Figure 1-4* illustrates this relationship of phase shift between two waves. Phase is important in the physical location of speakers — more about this later.

BASIC PRINCIPLES

Now let's look at a typical automobile sound system, some of its basic principles, and some of the improvements that you can make to it. The basic AM/FM radio or AM/FM radio/cassette provided by most automobile manufacturers is not physically and electrically capable of producing high-quality sound, especially deep bass.

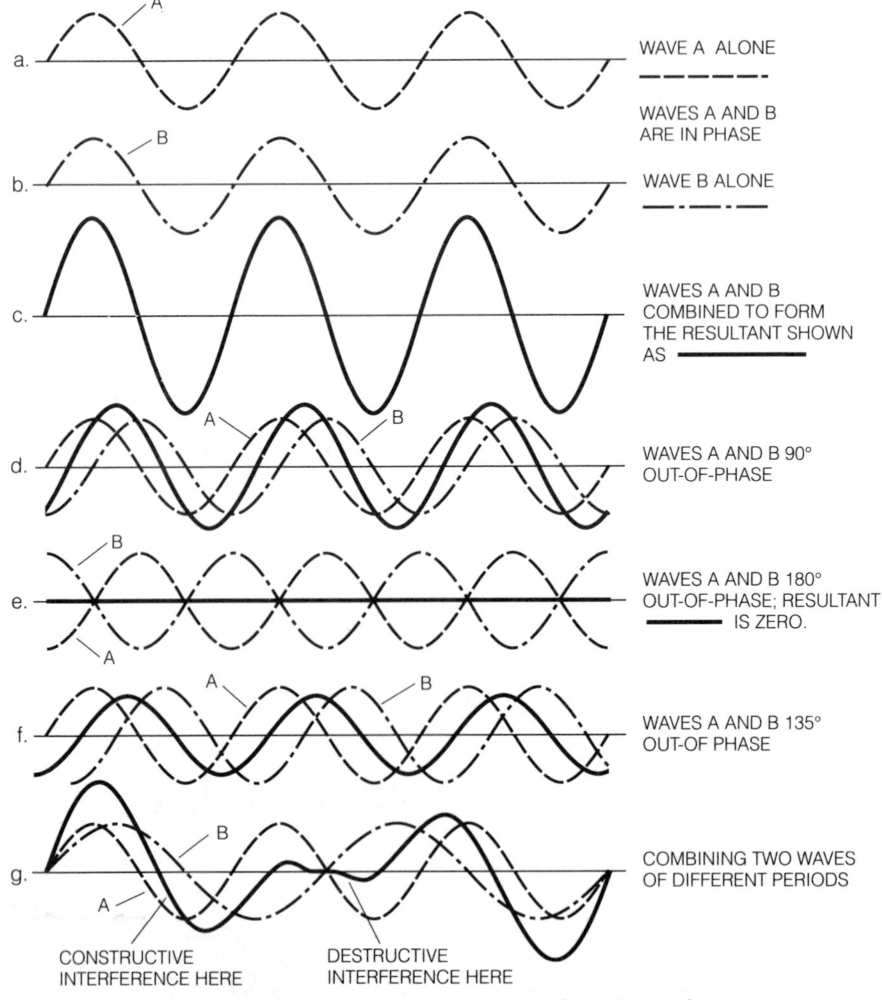

Figure 1-4. The concept of phase, phase shift and interference.

The basic system can be enhanced by replacing the weak links of the system. This means starting with the speakers since they have the most impact on the overall sound of your car stereo system. We will tell you the correct way to add speakers in Chapter 4.

Most factory radios can deliver only a few watts of output power per channel. To improve the sound to true high fidelity, you will need to add a good quality power amplifier. An amplifier is an electronic circuit that increases the amplitude of an electronic signal. It also is an important building block in a stereo system.

Figure 1-5 shows a complete automobile sound system with an AM/FM radio/cassette player, CD player, an equalizer feeding booster amplifiers for the front and rear speakers in the existing mounting holes, and a separate subwoofer amplifier and speakers for enhanced bass. The equalizer shown in the figure has a fader and a subwoofer crossover with selectable crossover frequencies. The fader lets you balance the output of front and rear speakers in a four-or-more speaker system.

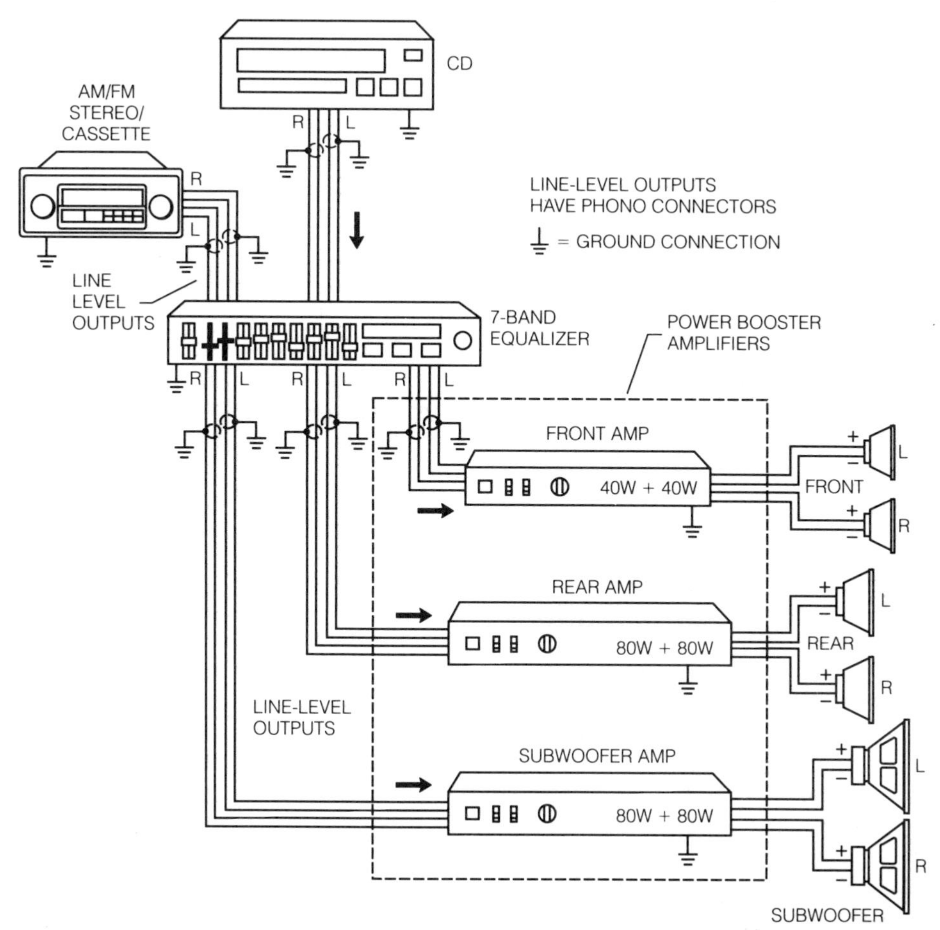

Figure 1-5. Complete automobile sound system with equalizer feeding booster amplifiers for front and rear speakers in existing mountings and a separate subwoofer amplifier for enhanced bass.

SYSTEM SPECIFICATIONS

Technology is advancing so rapidly that it sometimes seems like a sound system is out of date before it's even out of the box. Just a few years ago the maximum auto sound amplifier power was commonly 10 to 20 watts and the distortion was given at around 10 percent. These amplifier specifications are archaic when compared to those for modern amplifiers. Since the average person now spends many more hours inside an automobile, it's no wonder that more and more automobile owners are upgrading their systems so they can enjoy a higher quality of sonic realism.

To choose from the wide assortment of upgrade system components, you must know how they are rated. This means that you must understand the basic specifications (often called specs). In the remainder of this chapter, we will discuss the most common specs for sound systems: output power, distortion, and signal-to-noise ratio.

Output Power

The output power of an audio system is measured in electrical watts delivered to the speakers. The power is referred to as continuous average power and is simply calculated by Ohm's law using:

$$P = E \times I \quad \text{or} \quad P = I^2 \times R \quad \text{or} \quad P = \frac{E^2}{R}$$

where I is the RMS current delivered to the speaker, E is the RMS voltage across the speaker, and R is the resistance of the speaker or load. (See glossary for definition of RMS.)

When the output of a sound system is not loud enough, the normal reaction is to turn up the volume. If the system output power amplifier cannot deliver the power required, increasing the volume will overdrive the amplifier. Overdriving the amplifier causes distortion because the peaks of the signal are clipped or "flattened" rather than being a clean sine-wave shape. The square-topped waveform increases the high frequencies contained in the signal. As a result, a larger portion of the power output is used or directed to reproduce the higher frequencies.

Two things result—the first is distorted unpleasant sound; the second is possible damage to tweeters or high-range speakers. Once the amplifier begins clipping and the volume is further increased, a large portion of the energy is diverted to the tweeters until they can no longer handle the signal and are damaged or destroyed.

Power amplifiers not only must have the ability to deliver the overall power level required, but also must have the excess power capacity and dynamic range to handle the peaks of signal waveforms without distortion. Peaks in signal waveforms are just like instantaneous increases in amplifier volume and can cause waveform clipping if the dynamic range of the amplifier is not great enough. Specify your system power amplifier (and speakers to match) with some excess output capacity to avoid distortion and possible system damage.

Distortion

Distortion is an undesired change in the waveform of the original signal as it passes through an audio system. The result is unfaithful reproduction of the original signal. This means that the output signal is no longer just an amplified version of the input signal, but it has been changed in some way.

The five common types of distortion in amplifiers are:

- Frequency distortion
- Phase distortion
- Harmonic distortion
- Intermodulation distortion
- Transient intermodulation distortion

Frequency Distortion

Frequency distortion occurs when certain frequencies are amplified more than others. Capacitive and inductive effects in a circuit cause the distortion. The frequency response is the audio specification that reveals the degree of frequency distortion present in a circuit or system.

Sometimes frequency response is called frequency bandwidth. This is the difference between the high frequency (f_2), where the output power drops to one-half of its value at the middle frequencies, and the low frequency (f_1), where the power also drops to one-half its mid-frequency amplitude. These two frequencies (f_1 and f_2), called the -3dB (or "3dB down") frequencies, determine the portion of the 20-20,000Hz audible band over which the power is measured. (Some audio systems are rated over the range determined by the ± 1dB frequencies instead of ± 3dB.)

Most audio reproduction units (amplifiers, radios, tape players, etc) have a frequency response specification. An amplifier with a frequency response specification of 20-20,000Hz is better than one with 40-15,000Hz; that is, the wider the frequency bandwidth the better. Look again at *Figure 1-3* and notice that many instruments, particularly the string variety, span a wide frequency range, and require a sound system that accurately reproduces this wide range of frequencies.

Phase Distortion

Phase distortion occurs when one frequency component of a complex input signal takes longer to pass through an amplifier or system than another frequency. Though their amplitudes may be equally amplified (that is, no frequency distortion), differential phase shift may occur for the individual frequency components, therefore, the composite signal is changed or distorted.

Total Harmonic Distortion

Total harmonic distortion (THD) is probably the most common distortion. The percentage of THD depends upon the magnitude of the generated harmonics. As we discussed earlier, a sine wave signal is absolutely pure; that is, it has only one fundamental frequency component and no harmonic content. A harmonic is an integer (whole number) multiple of the fundamental frequency. For example, if the fundamental frequency is 500Hz, the second harmonic would be 1000Hz, the third harmonic would be 1500Hz, and so forth. Anything that happens to a sine wave signal that acts to change its waveshape will add other signal components (harmonics) to the fundamental frequency signal. Thus, if an audio amplifier modifies the shape of the waveform that it processes, the resulting harmonic signals will change the character of the sound produced by the speaker.

Harmonic distortion, sometimes less accurately referred to as amplitude distortion or non-linear distortion, occurs when the amplifying device is operated in a nonlinear condition. Therefore, an amplifier should operate in its linear region to prevent distortion of the signal that it is amplifying. The most common cause of harmonic distortion is the clipping of a signal when its amplitude is too high, such as when the volume control is turned up very high.

The percentage of harmonic distortion depends on the magnitude of the generated harmonics — the lower the number, the better. Of course, 0% distortion is the ideal condition. You must check distortion at several power levels and at several audio frequencies to make certain that the total harmonic distortion of an amplifier meets its specification.

Intermodulation Distortion

Intermodulation (IM) distortion occurs when combinations of new frequencies are generated by modulation within the amplifier or system. These new frequencies are equal to the sums and differences of integral multiples of the component frequencies of a complex wave. A true audiophile should learn how to interpret the results of this measurement. The book, *Making Sense of Sound*, referred to previously, contains more information on the measurement details.

Transient Intermodulation Distortion

Transient intermodulation (TIM) distortion occurs principally during loud, high-frequency music passages in solid-state amplifiers that use large amounts of negative feedback to improve frequency response and reduce harmonic distortion. Unlike THD and IM, TIM is not easy to measure, and the industry has not established a universal measurement standard.

Signal-to-Noise Ratio

Noise is any unwanted disturbance within an electrical or mechanical system that modifies the desired signal. To the listener of a sound system, this noise often manifests itself as static or hum. It may be a low-level background noise or an occasional burst of static that is only irritating, or it may be of such high amplitude and continuity that it obliterates the program material.

In measuring the quality of an electronic system, or in comparing one system with another, it is common to express the signal power P_s and the noise power P_n as a ratio. This signal-to noise (S/N) ratio of power levels is expressed in dB by the equation:

$$\frac{S}{N}\text{ (dB)} = 10 \log \frac{P_s}{P_n}$$

The same performance also can be expressed as a ratio of voltage levels by the equation:

$$\frac{S}{N}\text{ (dB)} = 20 \log \frac{V_s}{V_n}$$

Tape decks with Dolby® noise reduction systems have much better signal-to-noise ratios then decks without Dolby.

SUMMARY

With this look at some of the terms, definitions and specifications of sound systems in general, let's now look at some of the specific system components used in automobiles and consider some of their features.

System Components and Features 2

In this chapter, we focus on the different enhancements and additions that you can make to your system to achieve your personal listening criteria. Whether your requirements demand a complete system overhaul or just a simple radio head-unit change-out, the information presented in this chapter will help you choose the appropriate modifications.

RADIO

The radio is often called the "head-unit." It is the first, and probably the most important, choice that you must make. With today's technology, many features and sound controls are squeezed into a 2″ × 6″ × 6″ container. When selecting your tuner (the AM/FM receiver portion of the radio), you will find that analog (pointer along a scale) tuners are practically a thing of the past. Today, most radios have electronic digital tuners, which help to eliminate station drift. Electronic tuners also have either a liquid crystal display (LCD) or a light-emitting diode (LED) display. The display shows the station frequency in numerals, which simplifies tuning in a station. Most head units also offer the convenience of displaying a built-in time-of-day clock.

Usually the first choice for a head-unit feature is the number of station frequencies that can be stored in memory (presets) you want. Most electronic tuners have a minimum of five AM and five FM presets — some have up to 30 — which provide fast and easy tuning to your favorite radio stations.

An option is an FM signal booster. This compact external addition was designed with the traveler in mind. It is connected in-line between the antenna and the head-unit. It usually offers a boost of about 10dB for FM signals, which allows you to hear clearly stations a greater distance from the transmitter. *Figure 2-1* shows an example of an inexpensive FM signal booster.

Next, you must choose between the DIN-C (traditional double-shaft) style, as shown in *Figure 2-2*, or the DIN-E style (referred to as flat-face), as shown in *Figure 2-3*. You need to refer to your vehicle's radio chassis compatibility guide to help you make this decision. You can find information in your vehicle owner's manual or in an installation conversion guide that is available at most radio dealers. A relatively new option on some DIN-E style radios is the "pull-out" feature. This allows you to remove the head-unit and take it with you when the automobile is unattended. This is a proven security asset because a thief cannot steal what is not there. *Figure 2-4* shows an example of a removable chassis radio.

Figure 2-1. An inexpensive FM signal booster. *(Courtesy of Radio Shack)*

Figure 2-2. A DIN-C style (traditional double-shaft) radio chassis. *(Courtesy of Radio Shack)*

Figure 2-3. A DIN-E style (flat-face) radio chassis. *(Courtesy of Alpine)*

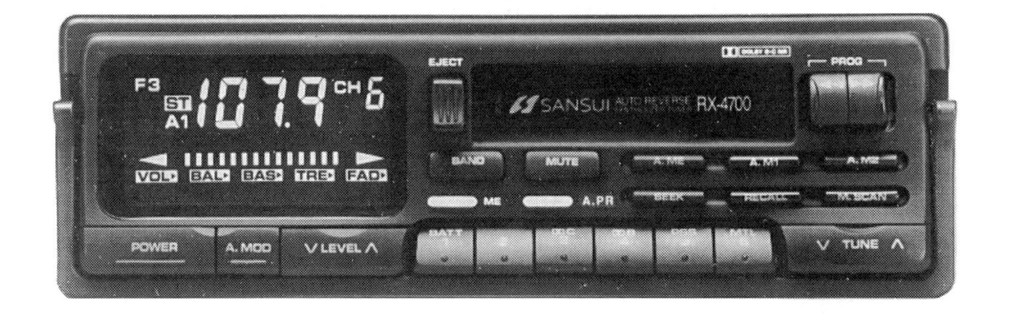

Figure 2-4. A DIN-E style removable-chassis radio. *(Courtesy of Sansui)*

Many other creature comforts are available and you may consider some of them essential. Most electronic tuners have seek and scan that search the band for available radio stations. Others offer memory scan that, with the push of a button, causes the radio to play about five seconds of each of your preset radio stations until you stop it.

If an equalizer is not in your system plans, then you should choose a head-unit with separate bass and treble controls. These permit you to adjust the tone to your preference more accurately than the single tone control.

The output power of the head-unit must be compatible with the speakers that are used in your system. Power output of the various radios ranges from about 2 watts to 20 watts per channel. If four speakers are in your design, then you will probably want a built-in fader to let you adjust the sound level ratio between the front pair and rear pair of speakers.

CASSETTE TAPE PLAYER

An excellent way to enjoy prerecorded music is to use audio cassette tapes. Most of the head-units on the market today have a built-in audio cassette tape player. The auto-reverse function is a popular option. This feature allows automatic, continuous play of both sides of a cassette tape, or you can control switching from one side to the other at any time by pushing a button. If a cassette player does not have the auto-reverse feature, it usually has an auto-stop function. This simply stops the cassette playing at the end of the tape on one side and disengages the tape transport. Most players have locking fast-forward and rewind buttons, which permit you to concentrate on driving while the tape is moving.

Another option usually included in better cassette players is the "key-off release." This function automatically turns off the cassette player and disengages the head and capstan; or a rarer feature, "key-off eject," in addition, ejects the cassette when you turn off the ignition switch. These features prevent the tape from being damaged, particularly in hot weather, by the tension placed on the tape when the tape transport is engaged. If your cassette player does not have either of these features, be sure to eject your tape manually before turning off the cassette player.

Some higher quality cassette units have three other options that the serious listener can consider. First, there is the Dolby noise reduction system, which you can usually turn on or off with a switch. This system is effective only when playing cassette tapes; it has no effect on tuner functions. It uses a filtering system that eliminates most of the annoying tape hiss. Next, automatic music search is a very convenient feature. When activated, it will automatically scan the cassette tape and stop at the beginning of each new song. Finally, metal or chrome tape capability is required to play the high-quality recordings made on metal tape. It should be played on a tape player designed for metal tape, otherwise, the tape player head will wear faster. *Figure 2-5* shows an audio cassette player that has many of the options we have discussed.

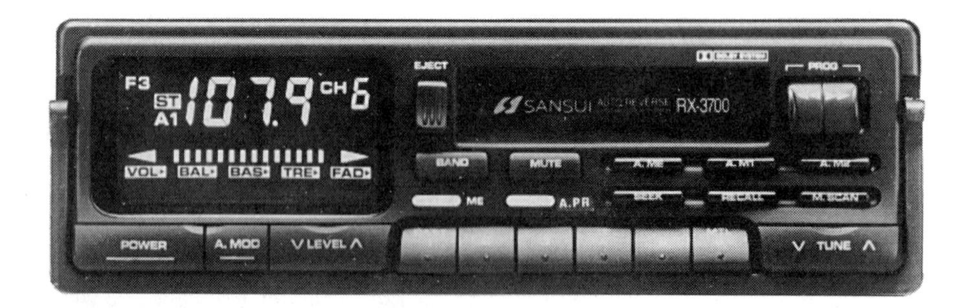

Figure 2-5. A cassette player that has many of the features discussed. *(Courtesy of Sansui)*

COMPACT DISC PLAYER

The compact disc player provides a clear and nearly flawless way to enjoy prerecorded music. Because the compact disc offers the superior digital reproduction of sound rather than the lower quality analog reproduction, it is virtually impossible to achieve equal sound reproduction through cassette tapes. (However, an advantage of cassette tapes is that you can make home recordings.)

The compact disc format provides immediate access to specific tracks on the compact disc. Most compact disc players, both home or auto, feature pause for mid-track interruption. Music searching is achieved through both fast-forward and reverse track locating which simply seeks the beginning or ending of a particular song. Also, most have automatic music search which simply fast-forwards or reverses internally through a song. Most compact disc players will play both the 5-inch and the 3-inch discs.

Many home compact disc players feature 8X oversampling, but most automobile compact disc players use only 4X oversampling. Oversampling is a digital filtering system that reduces digital noise levels to produce a much better sound. Some people believe that oversampling is directly related to the annoying skipping that sometimes happens when playing a compact disc; however, this is not true.

Figure 2-6 presents an example of an excellent high-power (36-watt), under-the-dash, compact disc player that has separate bass and treble controls, line-level output connections for an external amplifier, and 4X oversampling.

Another way to add a compact disc player to your system is by first choosing a head-unit with external compact disc input connections. *Figure 2-7* shows an example of an AM/FM radio and cassette player with an external compact disc input.

Figure 2-6. A high-power, under-the-dash, compact disc player.
(Courtesy of Radio Shack)

Figure 2-7. An AM/FM cassette player with external compact disc input jack.
(Courtesy of Panasonic)

To utilize this external input, you must have a portable compact disc player that uses either an external dc power source (such as through a cigarette lighter cord) or has internal batteries. *Figure 2-8* shows a full-featured portable compact disc player that has a one-key remote control and plays both 3" and 5" CDs.

As you can see, it is relatively easy to add a compact disc player, and your effort will be rewarded with a high-quality source of prerecorded music.

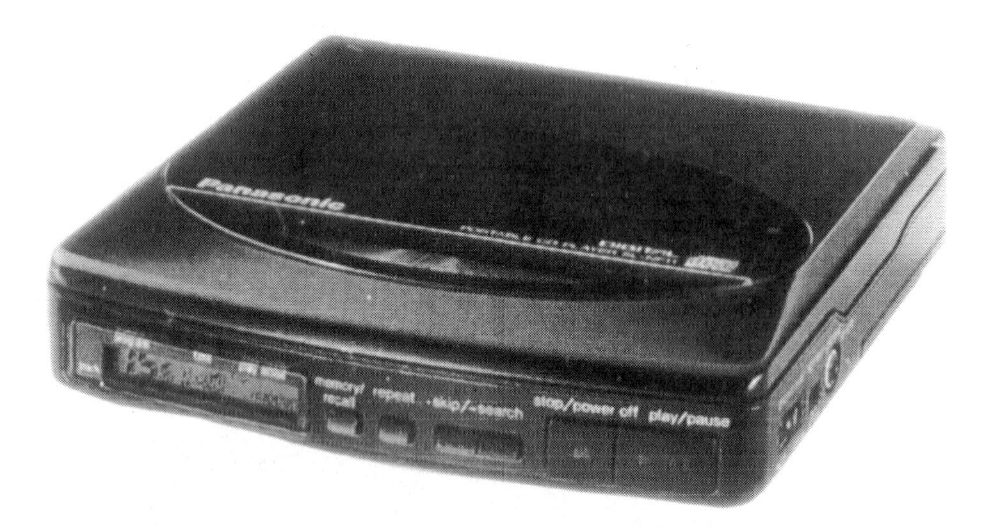

Figure 2-8. A full-featured, portable compact disc (CD) player.
(Courtesy of Panasonic)

EQUALIZERS

Over the past ten years, automobile equalizers have gone from practically non-existent to a bountiful line of choices. The primary function of an equalizer is to permit you to customize the sound to your taste or to correct sound deficiencies due to vehicle interior layout.

An equalizer utilizes boost and cut adjustments in each of several frequency bands to tailor the sound. Most automobile equalizers on the market use from five to ten bands, and the amount of boost or cut ranges up to 12dB.

The bands usually begin at 60Hz and go all the way up to 15kHz (15,000Hz). You can control the amount of boost or cut with slide-action controls, as shown in *Figure 2-9*, or with electronic level controls as shown in *Figure 2-10*.

If you want more power than that supplied by your head-unit, then an equalizer/booster may be the path for you. Several available equalizers have built-in power amplifiers to drive the speakers with 40 to 100 watts of signal power that has been "frequency-shaped" to suit your taste or match your vehicle interior. These equalizer/power boosters can either increase or cut the frequency response as much as 12dB in seven bands. Adding an equalizer/power booster may be the simplest method to improve the sound of your present car stereo system. *Figure 2-11* shows an example of an equalizer that has 40 watts of built-in power, fader control, 20-LED power meter and portable CD input jack.

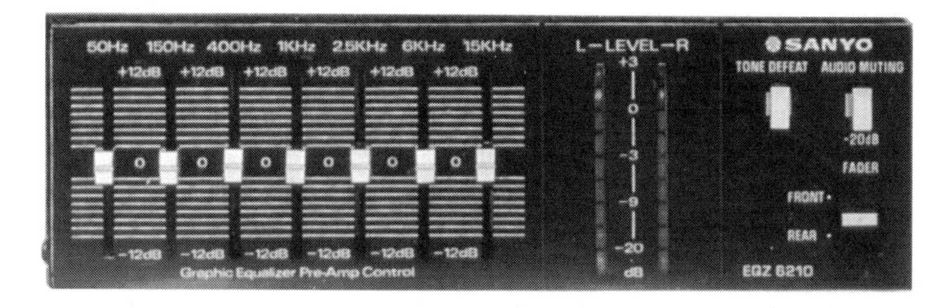

Figure 2-9. An equalizer with slide-action controls. *(Courtesy of Sanyo)*

Figure 2-10. An equalizer with push-button electronic controls. *(Courtesy of Alpine)*

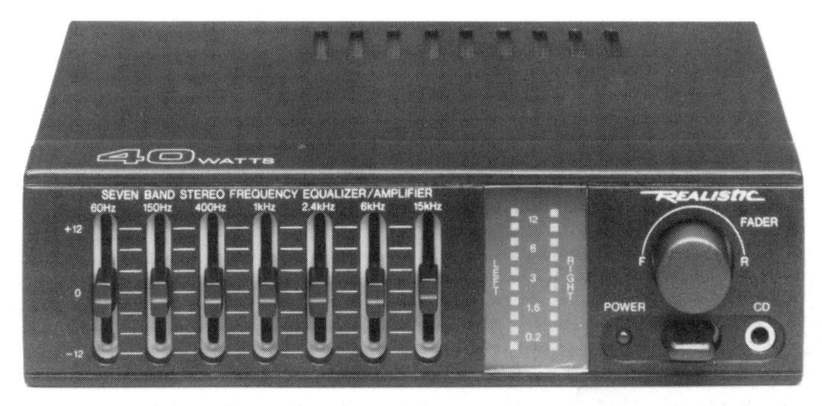

Figure 2-11. An equalizer with 40 watts of built-in power, fader control, 20-LED power indicator, and portable CD input jack. *(Courtesy of Radio Shack)*

Most equalizers on the market today offer four-channel operation, and provide a pair of powered front outputs and a pair of powered rear outputs. Your radio or head-unit may be equipped with a pair of "line level" or "pre-amp" outputs. The signals from these outputs have not been put through a power amplifier, so they provide a noticeably cleaner transition between the radio and the equalizer or power amplifier. Line-level connections are usually made by using a pair of shielded cables with phono-type connectors.

CAUTION

A pre-amp system can be used only if all components are line-level compatible. Connecting a speaker-level signal to a line-level input or output will permanently damage the unit.

Some equalizers offer various other options that some people think are necessary.

- Many have a built-in volume control, giving you two separate volume adjustments.
- A few have an external compact disc input jack.
- Some offer a "loud" switch that boosts the lower frequencies at low volume control settings so the music has a "full-bodied" sound at low listening levels.
- Some have a defeat button that locks out the equalizer adjustments so that the unit provides amplification with a flat frequency response.

For the serious system enthusiast, a passive equalizer is probably the way to go. A passive equalizer offers no internal amplification and is the cleanest form of tonal control. This type of equalizer has at its output pre-amplified (line-level) signals only. While proven to be very effective for many, most equalizer/boosters utilize a slightly "noisy" power amplifier. Because there is no power amplifier in a passive equalizer, it allows tonal adjustment of the signal and then passes the signal on to a better quality specified external power amplifier. *Figure 2-12* shows an example of a clean passive equalizer with phono-type line-level inputs and outputs.

A feature that is relatively new on the market is a built-in subwoofer crossover. You can use it to set up a multi-amplifier system with heavy bass subwoofer sound. An equalizer like the one shown in *Figure 2-13* can be a valuable addition to your system. This unit separates the signal for routing to as many as three different amplifiers (front, rear, subwoofer) at a very low noise level. It has a built-in fader and a continuously adjustable crossover point from 80Hz to 120Hz.

The built-in crossover has selectable frequency crossover points that allow you to pick between 60Hz and 100Hz, depending on the type of music you listen to. The switch is readily accessible on the front of the unit. This crossover system allows you to separate the line-level stereo signal coming from your radio unit three ways. First, a full-range pair of front outputs can be connected to your amplifier. (Remember, when utilizing a passive equalizer, an external amplifier is mandatory.) Next, a full-range set of rear outputs is available. Finally, a set of subwoofer outputs is available for a subwoofer amplifier. Remember, when we say "subwoofer amplifier" that this is not a specified amplifier. It is simply the amplifier you have designated to amplify the very low-frequency subwoofer signals coming from the crossover network.

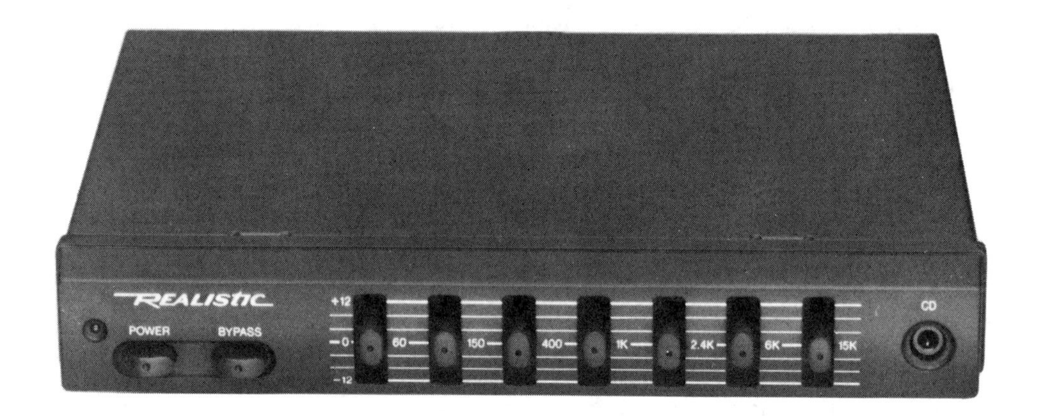

Figure 2-12. A passive equalizer with phono-type line-level inputs and outputs. *(Courtesy of Radio Shack)*

Figure 2-13. A passive equalizer with a continously adjustable crossover point. *(Courtesy of Alpine)*

If you are serious about the future of your sound system, but financially limited, then the addition of this equalizer is highly recommended. A good head-unit, equalizer, an amplifier and a single pair of speakers is all that is necessary to begin a great system. When your budget permits, you can add another amplifier and high-quality speakers. Finally, a pair of subwoofers and a power amplifier will complete the system. With a good solid plan in the beginning, you can use the building block method to accumulate a top-notch system.

POWER AMPLIFIERS

The popularity of convertibles, T-tops, sun roofs, sport utility vehicles, and cruising with the windows down while listening to your favorite music demands extra sound power to overcome noises from the outside environment. Since most factory-installed stereo systems are not capable of delivering more than a few watts per channel (2-5 watts), they are in dire need of a boost.

Unfortunately, it is a misconception to believe that upgrading the stereo head-unit will achieve your power needs. About 50% of the after-market improvements will add no more than 5 watts per channel. High-powered head-units, which are new to the market, can have up to 20 watts per channel. These radios make up about 30% of the market. The remaining 20% of the market may have no internal power amplifier at all. These radios are designed to work only with external power amplifiers and are usually the most expensive available.

Most currently available amplifiers are equipped with line-level inputs that have phono-type connectors. **(Remember, connecting a high-powered radio to this type of amplifier can permanently damage the amplifier and the radio.)** If you are going to work gradually up to a multi-amplifier system, but need a good place to start, then we recommend a good quality head-unit with built-in high-power amplification for immediate use, and line-level outputs for later addition of external amplification. Remember, when choosing an amplifier, you want to have as much power as possible without overpowering your speakers. To keep from causing permanent damage to the speakers, check the speaker specifications and do not exceed their maximum power-handling capacity.

Next, decide what your personal power needs are. If only a small boost is needed, then a small amplifier like that shown in *Figure 2-14* may be a good start. This amplifier uses speaker-level inputs (not pre-amp) for its source. Consequently, the final output might have a higher signal-to-noise ratio than one with line-level inputs.

Figure 2-14. A compact 20-watt per channel power booster.
(Courtesy of Radio Shack)

For the serious stereo enthusiast, we recommend one of the amplifiers shown in *Figures 2-15* and *2-16*. The unit in *Figure 2-15* is an 80-watt amplifier, thus it can deliver up to 40 watts per channel. This is plenty of power for a clean sound in many speakers. It can reproduce signals from 20Hz to 25,000Hz, so it can drive speakers from subwoofers to tweeters. With its noise-isolated input jacks, this amplifier can accept a pre-amp (line-level) signal and amplify it with minimal noise. The automatic power-on feature permits the unit to be mounted in areas that are not easily accessible. For safety, it has an overload protection device that protects the amplifier if a short occurs in the system.

Figure 2-15. A 40-watt per channel amplifier. *(Courtesy of Radio Shack)*

Figure 2-16. An 80-watt per channel amplifier. *(Courtesy of Alpine)*

The low-distortion bridgeable power amplifier in *Figure 2-16* is capable of a maximum of 100 watts per channel and reproduces sounds from 20Hz to 26,000Hz. Gold-plated terminals provide low-noise connections. This unit also has a PWM power supply, a 4-point adjustable crossover network, and acoustic compensation. This amplifier can deliver significant bass if it is connected in a subwoofer configuration.

These two amplifiers are watt-for-watt a pair of the best values we have found. They offer both the features and the power at a very reasonable price.

CROSSOVER NETWORKS

Most speakers are designed to reproduce a specific frequency range — woofer (20-3000Hz), midrange (1000-7000Hz), tweeter (3500-40,000Hz) — so a crossover network is necessary to route the proper frequencies to each speaker. The crossover network splits the full-range audio signal (20Hz-40,000Hz) into two or more ranges. The high frequencies are fed to the tweeters and the lows are fed to the woofers and midranges. Sending the wrong frequencies to the wrong drivers (speakers) can result not only in improper sound, but also can cause permanent damage to the drivers. When choosing your crossover, you must decide between a passive and an active crossover.

A passive crossover network utilizes resistors, capacitors, and inductors to form the filters. Capacitors allow high frequencies to pass while attenuating or blocking low frequencies. Inductors allow low frequencies to pass while attenuating or blocking high frequencies. Resistors attenuate all audio frequencies about the same and are used primarily for impedance or level matching. The capacitance and inductance values of the crossover components determine the crossover point. You must match the crossover frequencies to the useful frequency range of your speakers (woofers, midrange and tweeters.)

The passive crossover is installed between the amplifier and the speakers. Because of the installation point and components used, there is an inevitable power loss of up to 15 percent. *Figure 2-17* shows an example of a two-way passive crossover with a crossover point of 6000Hz.

You may prefer an active crossover network. It utilizes the same filtering methods as the passive crossover, but also has amplifying (active) circuitry. The active crossover has line-level inputs and outputs and is installed between the head-unit and the amplifier. Separating the frequencies at this level and amplifying the signal to overcome the loss of the filter components results in a higher power to your speakers. Remember, an external power amplifier is also used with an active crossover. Many equalizers have a built-in active subwoofer crossover with 2-way adjustable crossover points.

SPEAKERS

The speakers are probably the most critical elements in your stereo system. Without adequate speakers, it does not matter how powerful, expensive, or high-quality the other components in your system are. You can think of it as a funnel effect — no matter how large the sound going in, the speaker can reproduce only a certain amount of sound.

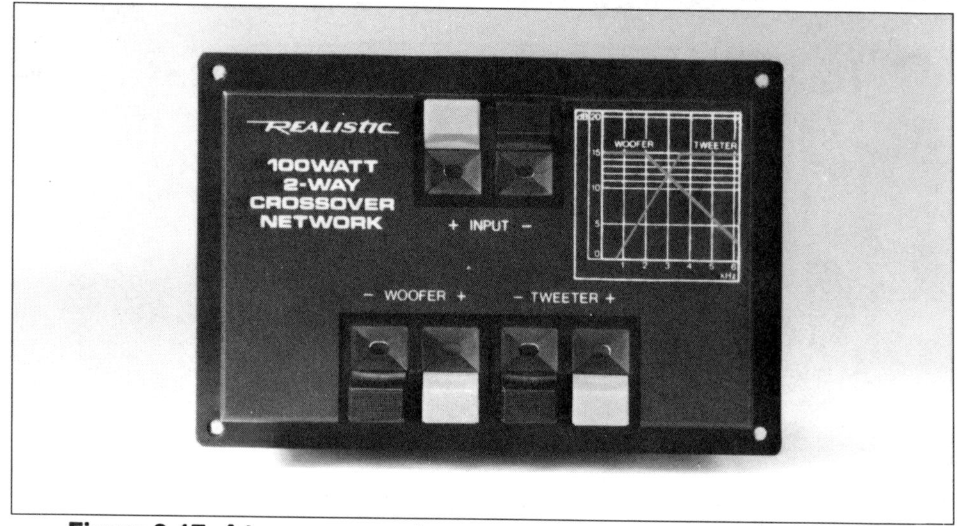

Figure 2-17. A two-way speaker crossover with 100 watts capacity.
(Courtesy of Radio Shack)

The best place to start improving your existing, factory-installed stereo system is by replacing the stock speakers with high-quality, after-market speakers of the same dimensions. The tonal quality and performance of most factory-installed speakers leave something to be desired because they are not designed with the avid sound enthusiast in mind. The factory speakers usually can handle only a few watts each.

The three commonly used types of speakers are woofer, mid-range and tweeter. The woofer is responsible for reproducing the lower tones of the listening spectrum (20Hz-3000Hz). Low tones require more air movement for sound, thus, the woofers are the largest speakers and require the most power. The mid-range reproduces the middle tones (1000Hz-7000Hz). The tweeter handles the higher frequencies (3.5kHz-40kHz) of the audio band. Progressively smaller amounts of power are needed from the woofer to the midrange to the tweeter.

To replace your stock speakers, first determine the location and size of speaker openings in your vehicle. You can usually obtain this information from your owner's manual, the vehicle manufacturer, or the speaker dealer's charts.

Once you determine the required speaker size, you need to decide the style of speaker you want. *Figure 2-18* shows a four-way, pre-enclosed surface mount speaker with 50 watts per speaker maximum power handling capacity. These work well when placed on top of a car's rear window deck, in the rear panels behind a pickup seat, or under the seat in a van or sport utility vehicle. There are also plate-mounted versions that have two or three speakers mounted on a rectangular plate. The plate can be bolted directly to the door or rear decks.

Another recent addition on the market is the "kicker" style box. *Figure 2-19* shows some excellent units. These boxes house 8-inch polypropylene woofers and are properly ported to reproduce thundering bass. They have top firing piezo tweeters to handle the upper-mid and high frequencies. These kickers can handle a maximum of 100 watts. Because of their compact enclosure design (approx. 12" X 18" X 6"), they will fit behind most pickup seats on the market today -- both full-size

Figure 2-18. A pre-enclosed surface-mount, bass-reflex design with woofers, midrange and tweeters. *(Courtesy of Radio Shack)*

Figure 2-19. "Kicker" style boxes. *(Courtesy of Kicker)*

and mini-truck body styles. They also work well in the rear cargo area of vans and sport utility vehicles.

We recommend these kickers because of their power handling capacity, tight construction, and ease of installation. Since these boxes are pre-fabricated, all you have to do is set them behind the seat, hook two speaker leads to each, and turn on the radio.

In subsequent chapters, we will describe enclosures that you can build yourself to make systems of your own. Such enclosures use raw drivers. A raw driver is a one-way speaker with no enclosure. It is intended to be used with a crossover and a raw driver tweeter. *Figure 2-20* shows an example of an 8-inch raw subwoofer driver. When this speaker is properly enclosured and used with a crossover and a good tweeter, such as the one shown in *Figure 2-21*, you can create a full-range system of your own.

Figure 2-20. An 8-inch raw subwoofer driver. *(Courtesy of Alpine)*

Figure 2-21. A good quality tweeter. *(Courtesy of Kicker)*

It is unwise to skimp on price, quality, or workmanship for speaker selection and installation. Poor choice of speakers and improper installation can prove to be the downfall of an otherwise outstanding system.

SECURITY

A security system is an invaluable addition to your automobile. With the rising frequency of automobile theft and vandalism, a theft deterrent is highly recommended. You definitely do not want to lose the stereo system you have worked so hard to assemble.

The first step in stereo security is to select a head-unit that is a "pull-out" style. Obviously, a thief cannot steal what is not there. Most radio dealers carry pull-out models with AM/FM stereos, cassette players, built-in high-power amplifiers and many other features that appeal to most enthusiasts. To install a pull-out model, you fasten the frame chassis permanently in the dash, then you slide the actual radio into the frame. When you leave the automobile, simply lift the handle on the front of the unit, pull the radio out, and carry it with you.

In addition to the pull-out head-unit, or instead of it, you should seriously consider protecting your vehicle and your system with an automobile alarm system. One of the better automobile alarms on the market today is shown in *Figure 2-22.* This system is loaded with features. The pager will beep you if an intruder tampers with your vehicle. A dual-function wireless remote control allows you to arm the alarm system after exiting. It also allows you to disarm the system for entrance to the vehicle. This remote control also has a panic mode that you can activate to sound the siren to scare off a potential thief. This panic mode also can be used in an emergency.

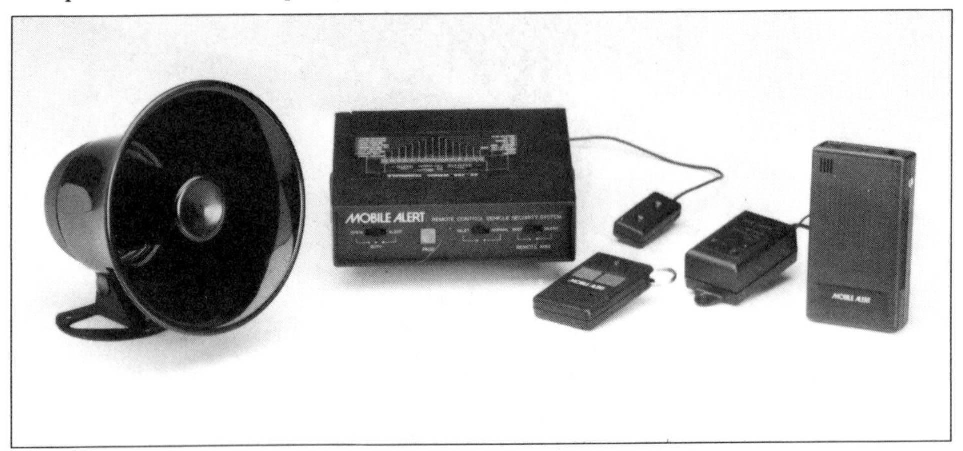

Figure 2-22. A vehicle alarm with remote pager. *(Courtesy of Radio Shack)*

Another feature that we have found to be quite valuable is the LED status indicator. With this small module mounted in an externally visible spot on your dash, many theft and vandalism attempts are avoided because the thief is pre-warned that an alarm is aboard.

We recommend that you add an electronic shock and motion sensor to this alarm system. Such a sensor (there must be motion) will detect someone trying to pry into a door or breaking out a window. Of particular importance, it can detect someone trying to jack up the automobile to steal a tire or to tow it away.

A fairly new type of alarm on the market is the infrasonic alarm. This alarm is mounted inside the cabin of the vehicle and senses sudden, sub-audible changes in air pressure, such as a car door or trunk opening. It utilizes a built-in, high-powered siren, and offers external siren hook-ups. It also has a two-function wireless remote for arm and disarm functions.

We cannot stress strongly enough how important some form of theft deterrent is. If quality stereo sound is important enough to spend the time and money to improve your sound system, then you should spend a little more to ensure that you, not a thief, enjoy it.

SUMMARY

You should now be more familiar with some of the terms used and the equipment available in the automotive sound industry. This knowledge will help you make better choices to obtain the sound system of your dreams.

STEPS FOR SYSTEM IMPROVEMENTS 3

There are many different ways to approach stereo system improvements. Your very first decision is to determine what you want to achieve. A big factor in that decision is whether you have no stereo equipment at all, or whether you are upgrading an existing system.

START FROM SCRATCH OR UPGRADE

Because of the wide variety of after-market stereo components and the automobile manufacturers' efforts to make lower-priced automobiles, many vehicles come with no radio at all. This is not as bad as it might seem because a factory-installed radio usually does not produce high-quality sound. Also, the car manufacturers usually charge too much money for factory radio systems and their upgrades. You can often take the same amount of money charged for the factory radio option and buy an aftermarket system that has more power and features and produces notably better-quality sound.

There are exceptions: A few of the higher-priced automobiles offer sound systems that pack a punch straight off the assembly line. Examples are the Infinity Systems, Inc. system from Chrysler, the Harmon-Kardon, Inc. JBL system from Ford, the Bose Corporation system from General Motors, the Lexus Nakamichi subwoofer/compact disc system and Acura's new DSP (digital signal processing) system which utilizes computerized sound tailoring. Because of the importance of automobile sound systems these days, other automobile manufacturers are beginning to concentrate more on their sound systems.

If your automobile is without audio equipment, you have a wide range of choices. You can install anything from a simple AM radio for listening to news, weather and low-fi music to a sound system that packs serious bass that can be felt as well as heard and produces true midrange and treble sounds.

If you already have a system and want to improve it, you may want to replace the head-unit (radio) and/or the speakers for the simplest and quickest upgrade. Or you may want to add a CD player, an equalizer, or a power amplifier.

PLANNING YOUR SYSTEM

Whatever you decide, start by making a system plan. Write down all the things you desire in your system. Your plan might include loudness, CD sound, awesome bass, and a heard-but-not-seen system. It is entirely up to you what is important. Also, include a list of the equipment you are removing so you can safely store it and reinstall the equipment when you sell your automobile. Once these decisions are made, then you must choose appropriate equipment to meet the criteria you have established. With your knowledge from the previous chapters, you should be able to understand equipment specifications and be able to make educated choices. Without a step-by-step plan, you will find that building a stereo system can turn into an overwhelming project.

ADDITION OR REPLACEMENT OF THE HEAD-UNIT

If your automobile has no head-unit at all, then it probably has a cover plate over the head-unit hole similar to the one shown in *Figure 3-1.* Even when a head-unit was not installed at the factory, the vehicle has factory speaker openings and the wiring harness usually includes the wires necessary for a head-unit installation. With no installed head-unit or speakers, you are actually one step closer to your completed stereo project before you ever begin because you need not take time to remove the existing equipment.

Figure 3-1. Factory-installed cover plate over hole for head-unit.

Removing the Cover Plate

The cover plate is usually just "snapped in." Examine the cover plate closely to make sure that no screws or bolts are holding it in place. If you find some, then remove them.

CAUTION

When prying against the vehicle's interior, wrap a soft cloth around the pry tool to help prevent scratching the vehicle's finish.

If there are no screws, then place a small flat-blade screwdriver or similar tool under the edge of the plate and slowly begin working it loose. On some models, you can put your hand under the dash and push from the back side of the plate until it pops out.

Locating Existing Power Wires

When the plate is removed, look for a wiring harness which is usually located behind the plate. It will have 5 to 20 wires in it. There should be no connections to the factory installed connectors. You must locate the existing wires that will be needed to power the new unit. You can do this by using a multitester, or even an inexpensive low-voltage test light. Both are available at electronics stores or most auto parts stores. The book, *VOM and DVM Multitesters,*[1] explains how to use a multitester. Begin with the wires by using a sharp knife or razor blade. Test the wires that are intended for connections to the head-unit that is to be installed. Connect the minus (-) probe of the multitester or one wire of the test light to chassis ground.

Now find a wire that supplies constant power from the battery with the ignition key off. To do this, turn the ignition key to the off position and begin touching the plus (+) probe of the multitester or the other wire of the test light to the available wires, one at a time. Make sure the multitester is set to measure a voltage greater than 12 volts. When the multitester indicates a voltage or the test light illuminates, you have found the wire you need. Write on a piece of paper the color and position of the wire.

Next, find an ignition key-switched power source. To locate this wire, turn the ignition key to the accessory position. Begin touching the remaining wires (do not include the wire already determined as the constant power source) until the multitester indicates or the test light comes on. Write down the color and position on your paper.

The constant battery power source serves two purposes. For radios, equalizers, CD players, etc. it provides an uninterrupted battery power to maintain clocks, timers and tuner memories with pre-programmed information. It also serves as the high-current power source for power amplifiers. However, for high-power independent amplifiers, a separate high-current power lead connected directly to the battery is recommended.

The ignition-switched power source also serves two purposes. The first is to control when power for full-circuit operation is applied to a component. The second is to energize a remotely controlled antenna or a switching circuit (relay) in a power amplifier that connects the high-current battery power to the power amplifier internal circuits. The high-current battery source remains isolated from the internal circuits (so the power amplifier is inoperable) until ignition-switched power is applied. In Radio Shack equipment, there is an orange wire that serves this switched power function for components coupled together. It triggers the switching circuit identified above.

Next, find a wire that has +12 volts applied when the headlights are turned on. This wire will be used to illuminate the head-unit display when the vehicle lights are turned on.

Running Your Own Power Wires

If you can't find the desired wires in the existing wiring, then you can run your own wires directly from the automobile's fuse block. The fuse block should have a terminal labeled "battery," that will supply constant power, a terminal labeled "radio,"

[1] *VOM and DVM Multitesters*, A.J. Evans, ©Copyright 1992, Master Publishing

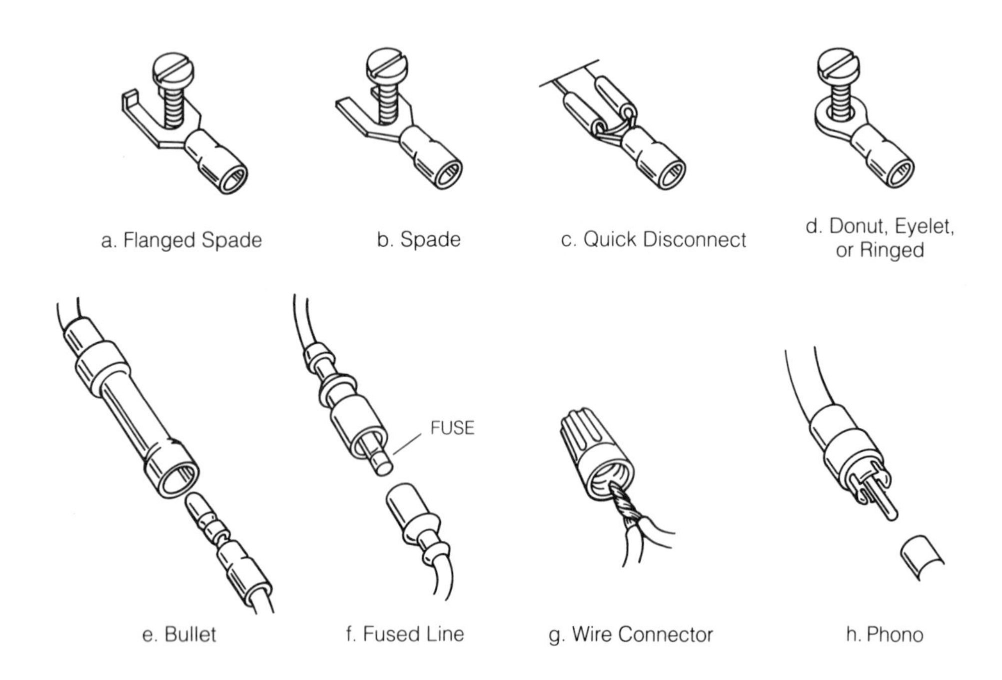

Figure 3-2. Terminal connectors.

"accessory," or "ignition," that will supply switched power and a terminal labeled "headlights" or "lights." Check these with your multitester or test light to be sure.

CAUTION

Before you start doing the actual wiring, disconnect the cable from the negative terminal of the battery.

You can use fuse block spade connectors (see *Figure 3-2b*) and relatively small (20 gauge) stranded copper wire, available from an electronics store, to run the wires to your head-unit compartment. Use different colored wires and note on your paper which is which to prevent confusion. If the present harness wires are very short, you can extend the wires so that they will be long enough to do all of the electrical connections to the head-unit end with the head-unit outside of its compartment. This will greatly improve the working space and eliminate the frustrations of trying to wire a re-mounted unit in a very limited area. Make sure all connections are insulated from each other so that shorted wires are not possible.

Establishing a Good Ground

Next, you must establish a good ground. You can use the existing ground wire, but it may be hard to find. It is often just as easy to run one yourself. To do this, find metal that is electrically connected to the vehicle chassis. Be careful, because on many newer cars, a nearby piece of metal may be isolated from ground by intervening plastic or other insulation and not provide a ground at all. *You must find metal with an uninterrupted path to the car's chassis.* Once you think you have located it, clip your minus (−) test probe of your multitester or your test light ground wire to the metal and the plus (+) probe of the multitester (set to measure 12 volts) or the other test light wire to the wire that you previously determined to be the constant battery source. If you have a multitester reading or your test light lights, then you have found a good ground. Use a different color of wire (preferably black) and a donut wire connection (see *Figure 3-2d*) to connect your new ground wire to the metal. If there is no available hole in the metal, then drill one. Take care not to drill into or through any vehicle components. Scrape away any paint from the area around the hole so a good electrical ground connection can be achieved. Once this is done, secure the donut connection tightly with a short steel screw or bolt and a lock washer to prevent it from vibrating loose.

Connecting the Head-Unit

Once the power wires are identified or established, connect them to the head-unit by simply following the included directions for your particular head-unit and wire it accordingly. Make sure that constant power and ignition-switched power to the head-unit are properly fused with 3A to 5A fuses. For many head-units, ignition-switched power is usually supplied through a fused line that feeds in through a 5-pin connector.

If you choose not to use a equalizer or amplifier, but just want to use the head-unit's basic output power directly to speakers, connect the speaker wires to the head-unit according to the manufacturer's included instructions. In most cases, manufacturers include interfacing connectors. If you are mixing manufacturers, you may need interfacing connectors to make up your own cables. Many electronics stores have quite a variety. Or you may choose to use crimped connectors as shown in *Figure 3-2e* and *Figure 3-3,* or wire nut connections as shown in *Figure 3-2g*. Be sure to observe polarity markings so the speakers will be properly phased. The speaker wires run in the harness by the automobile manufacturer are color coded. Use one color for the positive (+) output (usually green for left; brown for right) and the other color for the negative (-) output (usually white for left; gray for right). Refer to the installations shown in *Figures 3-6* and *3-7* for details on left and right and front and rear speakers. Once all wiring connections are made, plug in the external antenna cable.

We have assumed an antenna is in place. If one is required, see Chapter 7 for the installation of an antenna and its cable.

Mounting the Head-Unit

With all electrical connections complete, you are ready to mount the head-unit in place. Many automobiles require a mounting adapter kit. This can best be determined by consulting the salesperson where the unit is purchased. Your local auto parts store probably has adapter kits for many models. These kits come with explicit instruction for their particular application. Once the kit is installed and the wires and antenna connected, it should look similar to *Figure 3-3*, which shows an installation in a Chevrolet S-10 pickup. After following kit instructions and reinstalling any dash panels, your finished head-unit installation should look similar to the one shown in *Figure 3-4*.

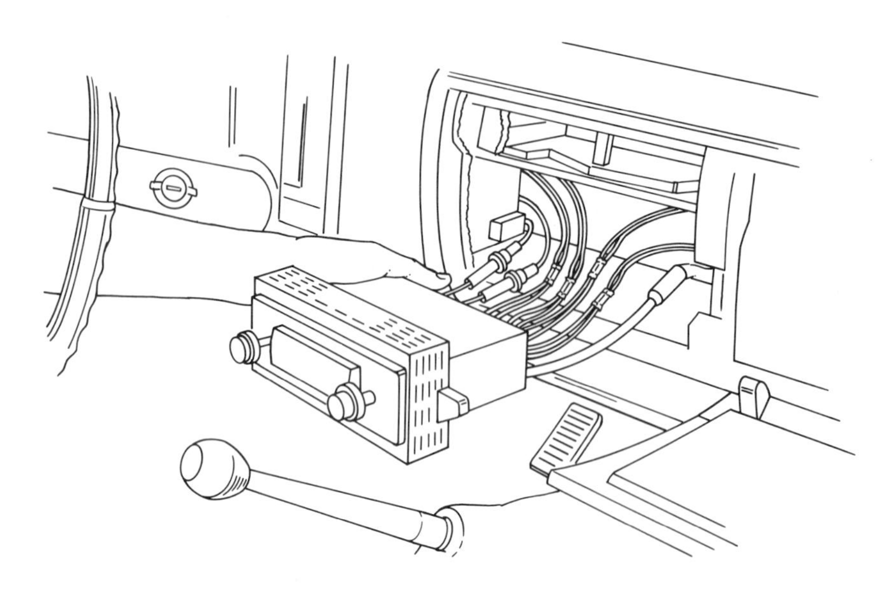

Figure 3-3. Head-unit wired and ready to be mounted.

Figure 3-4. Head-unit installation completed.

ADDING AN EQUALIZER/BOOSTER OR EQUALIZER

After installing your head-unit, you may chose to add an equalizer/booster to tailor the sound and increase the sound power. As we discussed in Chapter 2, an equalizer/booster is a good way to accomplish this.

CAUTION

When driving and enjoying your new sound system, you should not divert your attention away from your driving requirements to adjust the equalizer.

Equalizer/Booster

When mounting your equalizer/booster, try to pick an area easily accessible from the driver's seat. A neat looking installation can be achieved by using the included brackets for installation under the dash (See *Figure 3-5*). When searching for the spot to put the equalizer/booster, be sure to find a solid structure for the mounting. Many units include a template for drilling to ensure appropriate bolt hole alignment. If one is not included, mount the brackets to the equalizer/booster as shown in instructions that come with the unit, then hold the equalizer/booster in position and mark the places for the bracket holes. Select the correct size drill bit and carefully drill the holes. Take care not to drill into or through any wires or vehicle components. Bolt the unit into position as indicated in *Figure 3-5*. When it is in place, you are ready to begin wiring.

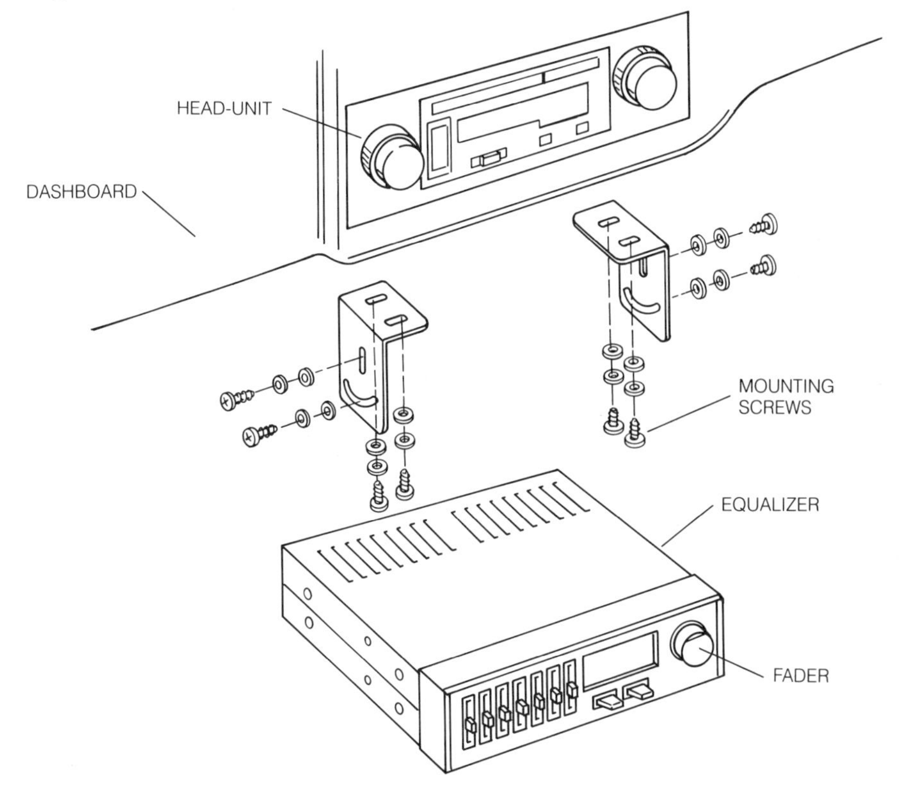

Figure 3-5. Mounting an equalizer under the dash.

CAUTION

Connecting the head-unit speaker-level outputs (which may deliver eight watts per channel or more of power) to the line-level inputs of an equalizer/booster can result in poor sound, distortion and/or even permanent damage to the equalizer/booster and head-unit.

Using Pre-amp Inputs

Wiring begins by finding a good ground. Use the one for the radio, if possible; otherwise, install a new one as discussed previously. The line-level outputs from the head-unit are connected to the equalizer pre-amp inputs with phono connectors (see *Figure 3-2h*). Prefabricated cables can be used or you can make your own. Since the unit is boosting power, the equalizer/booster drives the speakers. Connect the four speakers, being careful to wire them according to the proper front/rear, right/left configuration. Remember to keep the polarity correct. After all these connections are made, connect the ignition-switched and constant power leads. Make sure the power lines are fused properly.

The complete wiring should be similar to that shown in *Figure 3-6* where we have shown a head-unit and equalizer/booster interconnected in a pictorial view so that you can see the compatible connectors that are provided by the manufacturer. It also shows the interconnection of the orange control wire for the switching circuit. It also will help you decide the type of interface connector(s) you may need if you are mixing other manufacturers equipment. Some common connector-cable assemblies that may be needed are:

Item
18 ft. male 5-pin power and speaker cable
5-pin male universal cable
5-pin male cable for power and speaker connections
5-pin female auto stereo cable

Using Speaker-Level Inputs

When head-units do not have line-level outputs, the equalizer/booster must be driven with speaker-level outputs. The equalizer/booster must be equipped to accept speaker-level signals. The equalizer/booster of *Figure 3-6* will accept speaker-level signals. In fact, the cable harness that comes with the unit provides the flexibility of using either line-level or speaker-level inputs. When using line-level inputs, as shown in *Figure 3-6*, the 4-pin connector in the harness is left disconnected.

When the head-unit has no line-level outputs, the 4-pin connector is connected to feed the speaker-level signals to the input of the equalizer/booster. The signal path is from the head-unit speaker-level outputs, through the frequency modifications and amplification of the equalizer/booster, to the speakers. The speaker connections are the same as shown in *Figure 3-6.*

Equalizer

You may choose to use the power output from your head unit to drive the speakers directly, but you still want to have the adjustments from an equalizer. For such a case, you would just use a passive equalizer.

Now the signal path is from the head-unit line-level outputs, into the equalizer's line-level inputs, out the equalizer's line-level outputs, and into the head-unit's line-level inputs. The passive equalizer modifies the signal according to the frequency filter settings, feeds it back to the head-unit to be amplified and outputted to the speakers. There are switches that control the line-in and line-out connections on most head-units to indicate it is feeding an equalizer.

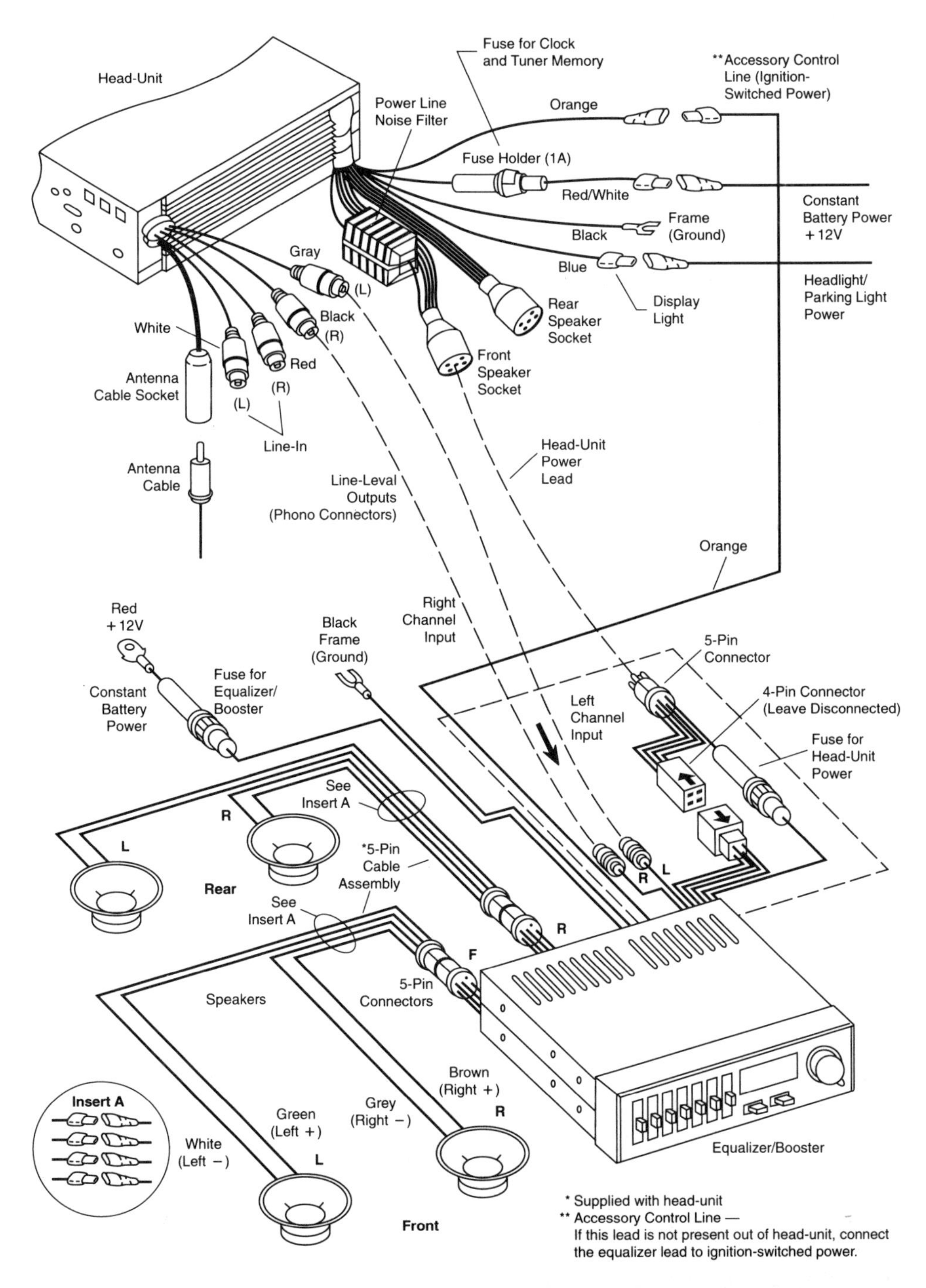

Figure 3-6. Pictorial wiring diagram using a head-unit, an equalizer/booster, and pre-amp inputs.

Wiring steps are the same as the pre-amp inputs. A pictorial wiring diagram using a passive equalizer is shown in *Figure 3-7*. Again, the compatible connectors provided by the manufacturer are shown. Interfacing to these connectors is required if mixing units from different manufacturers. When signal connections are complete, connect the ignition-switched and constant battery power leads. A proper ground connection is assumed.

After completion of the equalizer wiring and mounting, make sure all connections are insulated, and tuck excess wire up out of the way and out of sight. Use cable ties, if necessary, to hold the wires in place.

ADDING A POWER AMPLIFIER

If a louder and cleaner sound is among the features you've chosen, then a power amplifier can be the solution. Installation of a amplifier is a relatively simple task with a few precautions. It can be connected electrically between a head-unit and speakers, or between other components (e.g., an equalizer) and the speakers. Its input is a line-level input and its output is a speaker-level output. If your head-unit does not have line-level outputs, then you must use a amplifier with speaker-level inputs. This is not recommended because it produces a lower quality sound, and there is a good chance you will damage the amplifier unless the amplifier is designed to accept speaker-level inputs. If an amplifier is in your plans, we recommend that you purchase a head-unit with pre-amp line-level outputs.

Remember that automobile amplifiers are different from home stereo amplifiers in that they are designed to be powered by a 12-volt, negative-ground electrical system as is used in most automobiles today. Wiring your amplifier to any other electrical system will result in permanent damage to the unit.

CAUTION

Because of the high sound level that can be achieved with a power amplifier, you must use self-discipline and not play the radio so loud that you can't hear important traffic sounds, such as warning horn blasts and sirens. Also, continued exposure to high sound levels is known to cause hearing impairment.

Mounting the Amplifier

Unlike your radio installation, you will probably want to mount your amplifier before wiring. Because of the high power in an amplifier, it should be placed a *minimum* of three feet from the head-unit to reduce electrical interference. Recommended positions for the amplifier include the trunk, under the seat, or on the firewall above the passenger's feet.

CAUTION

Before drilling or screwing, check underneath the carpet and on the other side of the panel to prevent puncturing existing wires, gas lines, or hoses.

An amplifier can be mounted horizontally or vertically — it makes no difference — simply position it for optimum space utilization. Assure adequate ventilation to prevent amplifier overheating. Self-tapping screws are usually the simplest method for securing your amplifier. *Figure 3-8* shows how an amplifier has been secured to a solid surface underneath the rear deck inside a trunk.

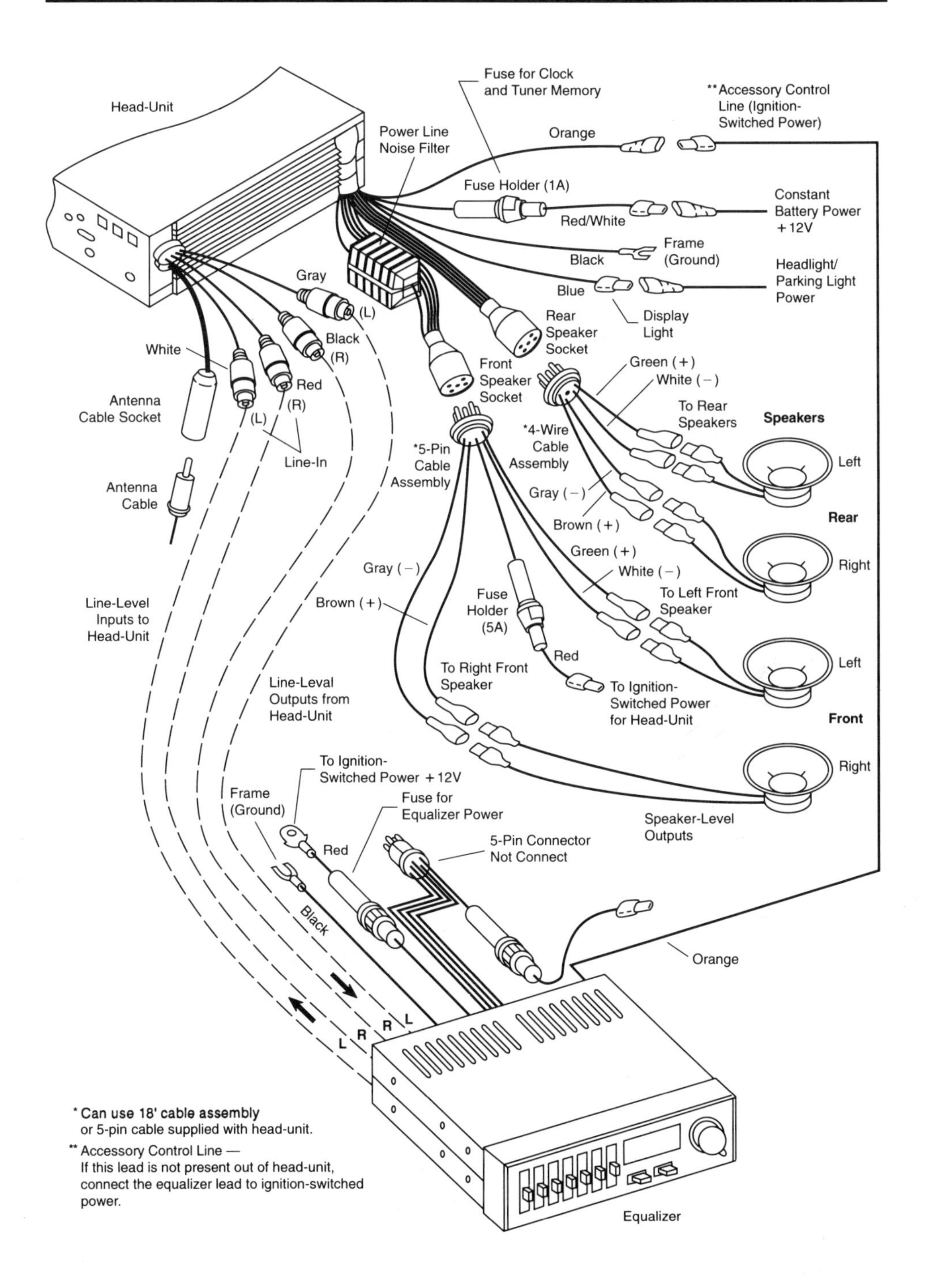

* Can use 18' cable assembly or 5-pin cable supplied with head-unit.

** Accessory Control Line —
If this lead is not present out of head-unit, connect the equalizer lead to ignition-switched power.

Figure 3-7. Pictorial wiring diagram using a head-unit and a passive equalizer.

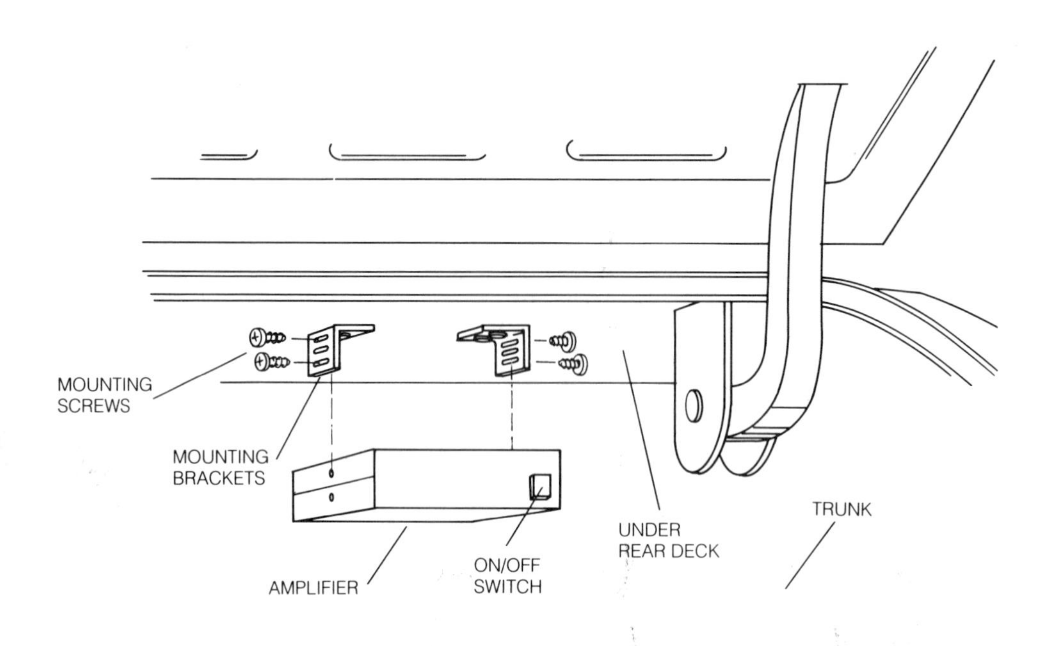

Figure 3-8. Mounting a power amplifier in a trunk.

Wiring the Amplifier

When the amplifier is mounted securely to the automobile, you are ready to wire it to its power source, speakers and input signal. *Figure 3-9* shows the required connections. Carefully read the instructions and observe the colors of the wires coming out of the amplifier. Find the amplifier ground wire and connect it to a good chassis ground. Because of the large current that the power amplifier requires, it is imperative that the ground wire be connected properly, as discussed previously, to provide noise-free operation. Check it with a multitester or test light to make sure the ground wire is making bare metal contact with the vehicle chassis. It should be kept as short as possible and be connected tightly with a bolt or a sheet metal screw.

Next, connect the "turn-on" lead. It connects to a switch (relay) inside the amplifier that automatically turns the amplifier on and off as the head-unit is turned on and off. It must be connected or the amplifier will not come on or the high-current battery power will not be internally connected. For the equipment in *Figures 3-6* and *3-7*, this is a special orange control lead sometimes called "remote on/off."

CAUTION

Do not connect the "turn-on" lead to a constant power source. If you do, the amplifier will be on continuously and run down your battery.

If your head-unit is equipped with a power antenna, the amplifier "turn-on" lead can be connected to the same lead that controls the antenna motor. If not, then connect it to an ignition-switched power source. Because this wire will carry very little current, a 20-gauge wire is adequate if you must lengthen the wire.

Next, connect the inputs to the amplifier. A majority of radios and amplifiers utilize phono plugs like those shown in *Figure 3-2h* for line-level connections. There are many varieties of cables with phone connectors and varying lengths from which to choose.

Connect the speaker leads to the amplifier as shown in *Figure 3- 9.* Note the color of the wires for the right-hand and left-hand speakers and note which is positive and negative for each speaker. Simply match the negative terminal of the speaker to the negative terminal of the amplifier, and positive terminal to positive terminal as shown in *Figure 3-9.* The speaker wires can be hidden under the carpet.

Finally, you must run the main power lead. This fused wire must be connected directly to the positive terminal of the automobile's battery. Use 10- or 12-gauge wire; a smaller wire (which is a larger gauge number) will limit the performance of your amplifier. Insert a 20-ampere in-line fuse between the battery and the amplifier, as close to the battery as possible, and yet in a place that will provide convenient access. Because this wire goes directly to the battery, it must pass through the firewall and into the engine compartment. Examine the firewall to see if you can find an existing hole to run the wire through. Sometimes an existing hole is camouflaged by a rubber plug. If you can't use an existing hole, locate an area where you can drill a hole, and

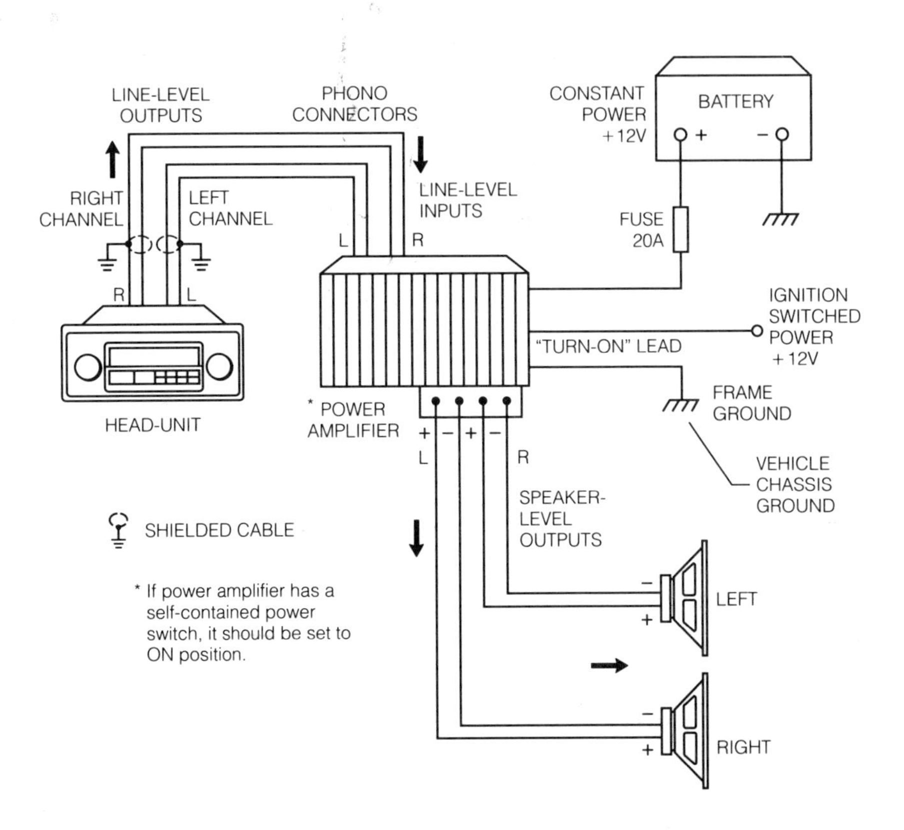

Figure 3-9. Wiring diagram for connecting a power amplifier

insert an appropriate rubber grommet in the hole to protect the wire. Carefully examine both sides of the firewall to be sure you won't drill through components such as brake lines, brake reservoir, electrical cables, etc. If the amplifier is mounted in the trunk or under a seat, the power wire can be run down the side of the automobile under the door threshold plate as shown in *Figure 3-10.*

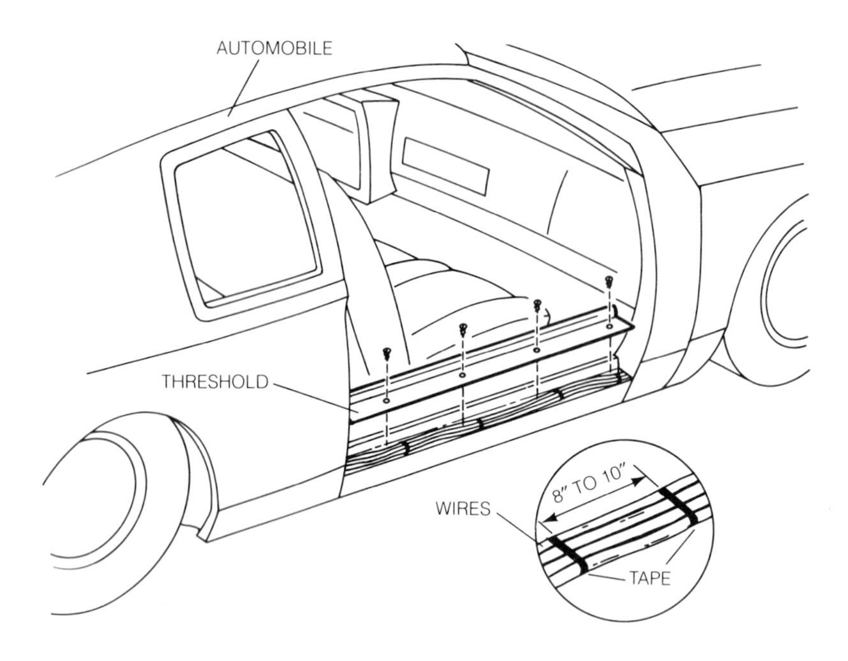

Figure 3-10. Routing wires under threshold plate from front to rear of an automobile.

CAUTION

Before connecting the main power, disconnect the negative terminal connector of the battery.

When all other wiring is completed, connect the main power wire to the positive battery terminal. Use the appropriate sized crimp-on ring connector which may be screwed under the positive terminal lead. Reconnect the negative battery terminal connector and enjoy your new audio system.

ADDING A CROSSOVER NETWORK

As we discussed before, a crossover network's function is to direct a certain range of audio frequencies to a particular speaker or speakers. A passive crossover is the more primitive of models on the market today. It uses a combination of passive low-pass and high-pass filters which is less efficient and costs less than an active crossover. Active electronic crossovers usually permit the crossover point to be adjusted to select the frequency range specified for the speakers to be used. They may provide greater flexibility, lower distortion, better transient response, and higher efficiency.

Passive Crossover

To install a passive crossover, simply connect the speaker output from the amplifier or head-unit to the crossover input. Then connect the speaker wires from the midrange speakers or tweeters to the high-frequency section output and the speaker wires from the woofers to the low-frequency section output as indicated in *Figure 3-11.* Power and ground leads are not necessary on a passive crossover.

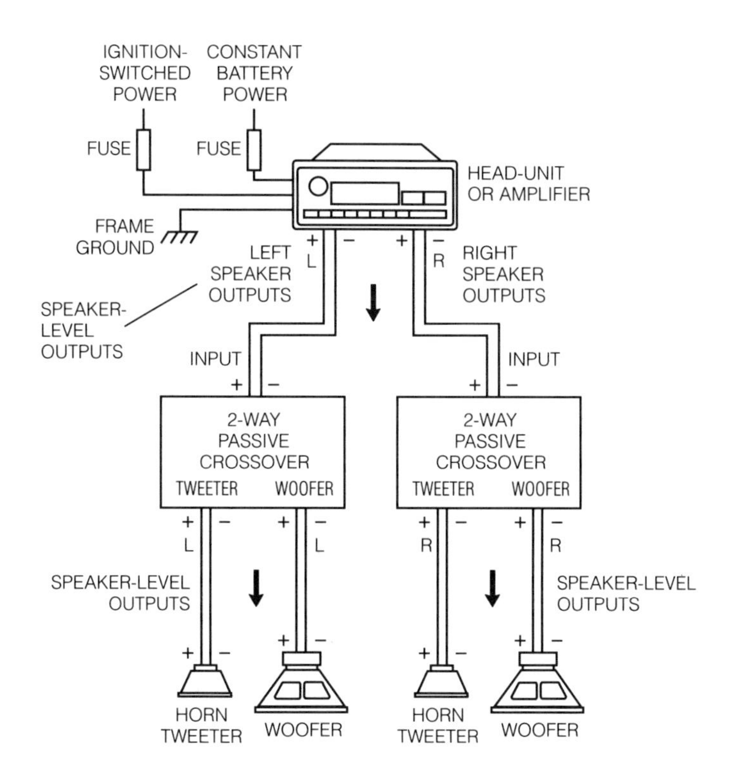

Figure 3-11. Wiring diagram for connecting a passive crossover.

Active Crossover

An electronic active crossover usually handles line-level signals, thus, phono connectors are normally used for the input and output connections as shown in *Figure 3-12.* The physical installation is much like that described above for the power amplifier because the unit does require ignition-switched power and a good signal ground.

ADDING SPEAKERS

When planning your system, give as much (or more) attention to the speakers as you do any other component. You must consider physical size, power-handling capability, and frequency response. We will cover actual speaker replacement in Chapter 4, but here are a few things to remember in choosing your speakers.

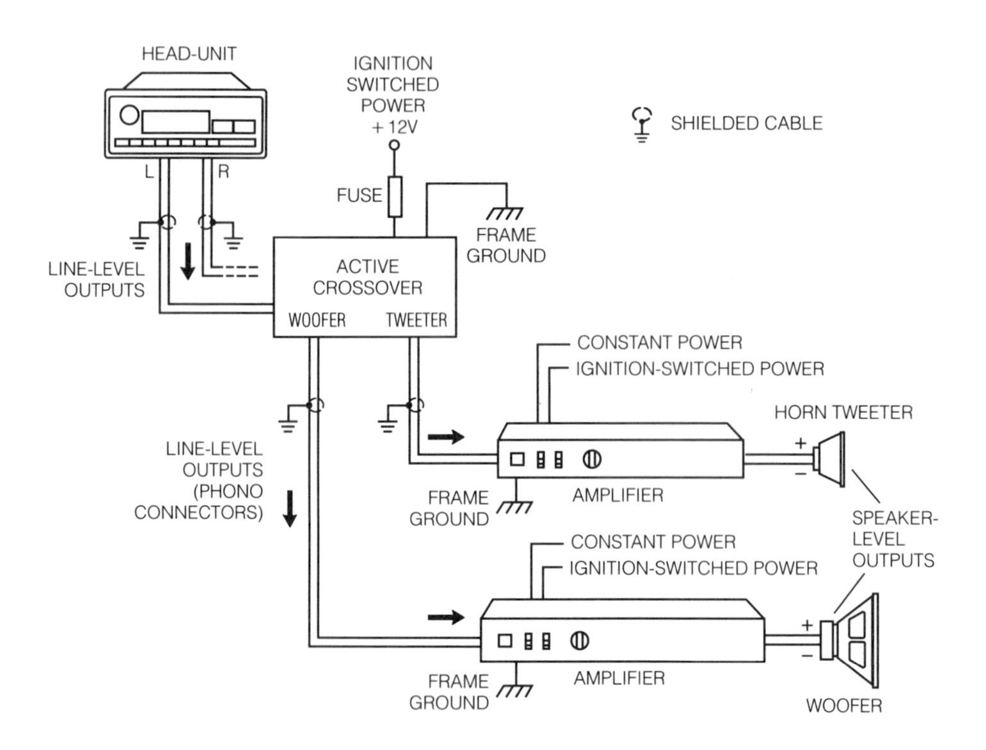

Figure 3-12. Wiring diagram for connecting an active crossover.

The speakers you choose must be capable of handling the sound power your audio equipment produces. Thus, you must decide on the power component (head-unit, equalizer/booster, or amplifier) and determine its maximum power output. Once this is decided, you must choose a speaker that has a maximum power handling capacity equal to or greater than the peak power output of the amplifying component. Most speaker specifications list the speaker's maximum or peak power handling capacity.

Of course, an alternative way to plan your system is to decide first on the maximum power and speaker size you want to use, then back up from there to choose the power amplifier required to drive the speakers, and finally, choose the sound source to drive the amplifiers.

SUMMARY

In conclusion, we cannot stress enough how important calm persistence is during your stereo building or modification project. Getting frustrated will not help, and could cause you to hurt yourself or damage an expensive part of your automobile. With a good plan, well-chosen audio equipment, proper tools, and competent work practices, you will have an outstanding automobile sound system.

REPLACING EXISTING SPEAKERS 4

Speakers are the final component in any sound system. They are one of the most important parts — if not *the* most important — of sound reproduction. Speakers have such a tremendous effect on the sound of an audio system that one poor speaker can ruin otherwise grand sound. The speakers in factory stereo systems usually are not capable of reproducing high-quality sound. By replacing them with upgraded speakers, you can enjoy the top-quality sound you would expect from a good home system.

Shopping for car speakers is not as simple as shopping for home speakers. If you were building a home speaker system, you could choose just about any size speaker available. The challenge of car audio systems, however, is to obtain the best possible sound using speakers that will fit in the vehicle. The intent of this chapter is to help you achieve this goal.

LOCATING FACTORY SPEAKER OPENINGS

The first step in your speaker replacement project is to locate the factory speaker openings. Manufacturers install factory speakers in many ways, but we will concentrate on the most common installations.

Figure 4-1 shows the most common locations for factory-installed speakers for coupe, sedan and hatchback automobiles. A coupe is a two-door automobile with a trunk, such as a Ford Thunderbird.® A sedan is a four-door vehicle with a trunk, such as a Chevrolet Caprice® Classic. A hatchback is a two-door automobile with a rear lift gate and no trunk, such as a Pontiac Firebird.®

Figure 4-2 shows common speaker opening locations on many pickups and sport utility vehicles. A sport utility vehicle is a two- or four-door vehicle with a rear cargo area and lift gate, such as a Ford Bronco® or Chevrolet Blazer.®

After referring to the speaker location diagrams in *Figure 4-1* and *Figure 4-2,* begin searching the automobile for the speakers. A sure clue to a speaker location is a grille covering as shown in *Figure 4-3.* Other times, they can be identified by small holes or slits in the auto's interior at the suggested locations. Some newer automobiles completely disguise the speaker by covering its location with a sound transparent material that matches the other interior materials in the automobile, as in the Pontiac Firebird's rear side panels shown in *Figure 4-4.* In this case, you must locate the speakers by ear — by listening for the source of the sound.

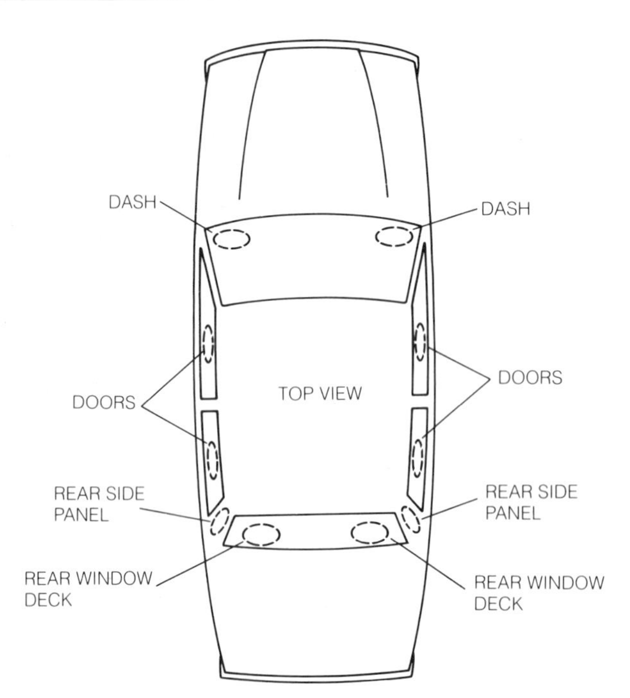

Figure 4-1. The most common existing speaker openings in an ordinary coupe, sedan and hatchback.

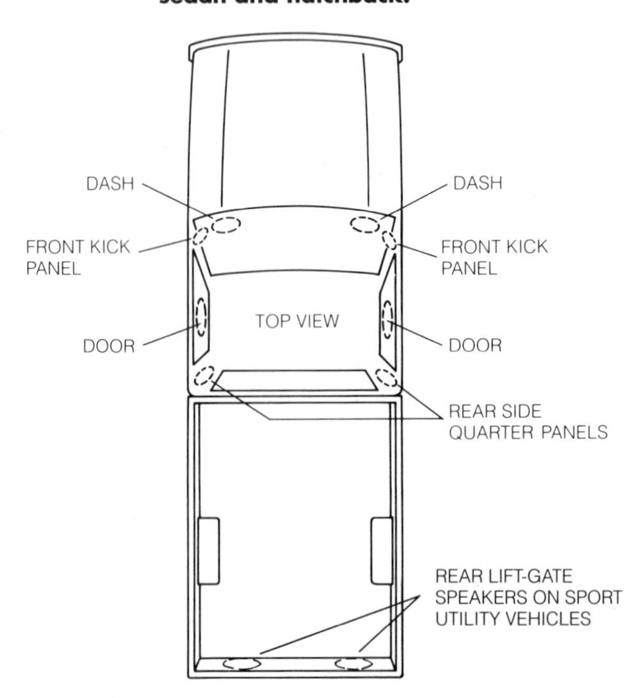

Figure 4-2. The most common existing speaker openings in an ordinary pickup and utility vehicle.

Figure 4-3. A grille covering sometimes can be used to located factory-installed speakers.

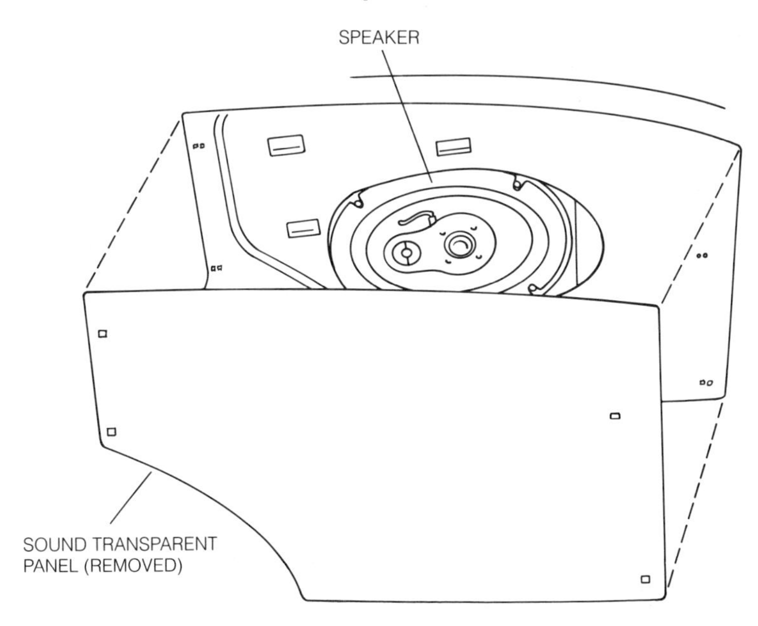

Figure 4-4. A speaker may be completely hidden when it is covered with a sound transparent material that matches the automobile's interior material.

REMOVING FACTORY SPEAKERS

After locating a speaker, you can begin removing the factory speaker. Many times, this is far simpler than you might expect.

Door and Side Panel Speakers

Removing a speaker using existing factory openings in a door or side panel is sometimes as simple as removing a cover and four screws and disconnecting the two speaker wires. Replacement is just as simple: connect the speaker wires to the new speaker, place it in the opening, secure it with screws, and replace the cover. This is illustrated in *Figure 4-5.* Although this method is most commonly used in door and side panel installations, it may be used almost anywhere. In some cases, the existing wire connectors may not match those needed for the new speakers. Simply cut off the old connectors and replace them with the correct matching connectors, either solderless (crimp) or ones that need to be soldered. Refer to *Figure 3-2* for the different connector types. Mark the wires and observe correct wiring polarity.

Rear Deck Speakers

Next, let's consider the speakers that are installed in openings in the rear window deck. These require a little more effort because they usually must be accessed through the trunk, at least for the electrical connections.

In some cases, the speaker was installed from the top into the rear deck opening and the speaker flange rests against the top of the rear deck. To remove the speaker, snap off the grille cover to reveal the bolts that hold the speaker in place. Remove the bolts, carefully lift the speaker, and disconnect the speaker wires. Simply reverse the procedure to install the replacement speaker.

In other cases, the speaker is suspended from the bottom of the rear deck by four bolts protruding downward from the rear deck. First, disconnect the speaker wires. Then support the speaker to keep it from falling while you remove the four nuts.

If the replacement speaker fits the opening and mounting holes, the installation is rather simple. Align the mounting holes in the speaker, push it in place, and hold it up while you start the nuts. Be careful not to puncture the cone of the new speaker on the

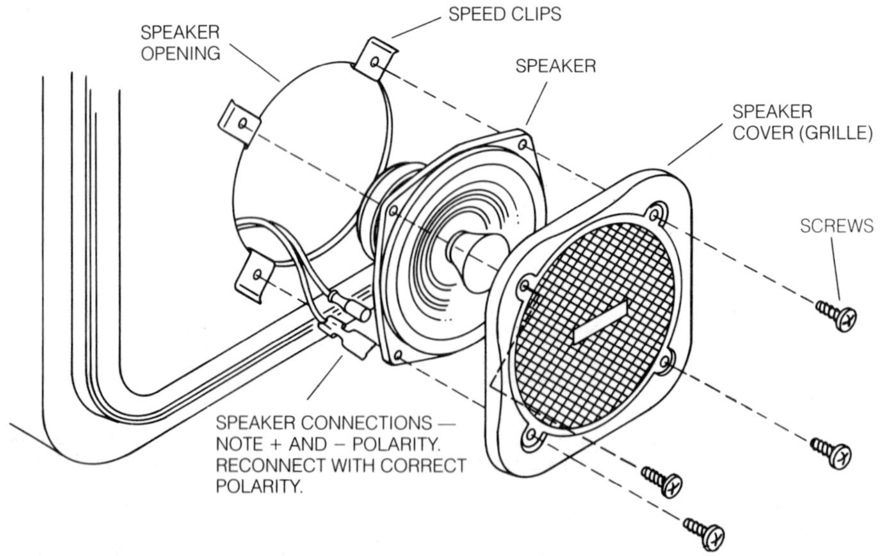

Figure 4-5. Replacing most factory speakers in doors and side panels is usually simple.

protruding bolts. Tighten the nuts securely but don't overtighten. Connect the speaker wires with correct polarity and you are in business.

In either of the above cases, if the replacement speaker does not fit the opening exactly and/or new mounting holes are required, place the new speaker upside down on top of the rear deck and mark the hole locations. Examine the area under the rear deck where the new holes will be drilled to make sure that nothing will be damaged. Drill the holes, insert the bolts, mount the speaker and tighten the nuts. If the opening is too small for the replacement speaker, enlarge it with a saw or rasp. Details for cutting an opening are given later in this chapter and also in Chapter 5.

Another method commonly used by manufacturers in rear deck installations is a swing bracket. A typical one is shown in *Figure 4-6.* These swing brackets are used mostly on newer model automobiles. Usually, the bracket is hinged on one side and has a quick-release latch on the other side. The spring pressure of the bracket holds the speaker tightly against the bottom of the rear deck. Keep in mind that only a speaker of almost identical size can be used correctly in this bracket. You may want to remove the factory speaker and take it with you when choosing its replacement. To remove the existing speaker, support the speaker while you release the latch. Carefully lower the speaker and disconnect the speaker wires. Simply reverse the procedure to install the replacement speaker.

Dashboard Speakers

Probably the most difficult place that factory speakers are installed is in the dashboard. Of the cars and trucks produced in the last ten years, 90% have factory speaker openings somewhere in the dash. These can usually be identified by a grille or a pattern of small holes. Study closely the area around the speaker grille. Sometimes hidden screws hold the grille assembly in place. Locating these hidden screws can save you unnecessary work. Unfortunately, most dash grilles cannot be removed, and

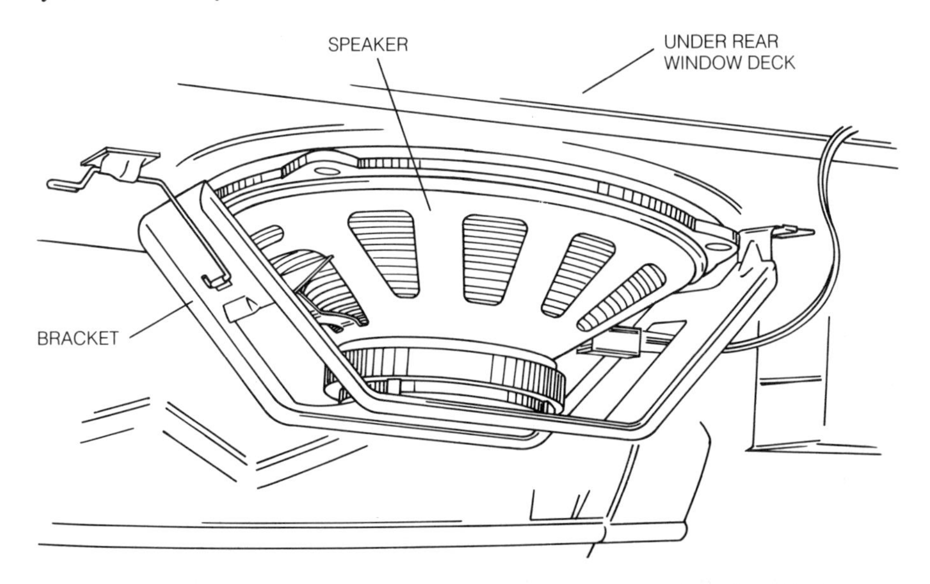

Figure 4-6. A swing bracket is sometimes used in rear deck installations.

if you try to, you can permanently damage them or the dash. In these cases, you will probably have to remove the complete dashboard assembly.

Once you have decided that dash removal is necessary, you will need to determine how it is secured. While many dashboards are made in one piece, some are in sections. To avoid needless work, remove only the necessary sections. To hold the dash, manufacturers usually use screws which are often hidden under the front lip of the dash. Another place to look is at the point where the dash meets the windshield. If you can't find screws in these areas, then, to avoid unnecessary damage, you should consult a maintenance manual or a professional for assistance with your specific make and model automobile.

Once you have removed all screws from the dash, you can begin carefully removing it. This is a two-person job, so enlisting the help of a friend will be quite beneficial. When the dash is out, you should be able to see the speakers as shown in *Figure 4-7.* Remove the factory speakers and install the new ones using one of the methods discussed above. A right-angle screwdriver and/or a hand ratchet with deep sockets can be a real help in the space-limited area on the dash.

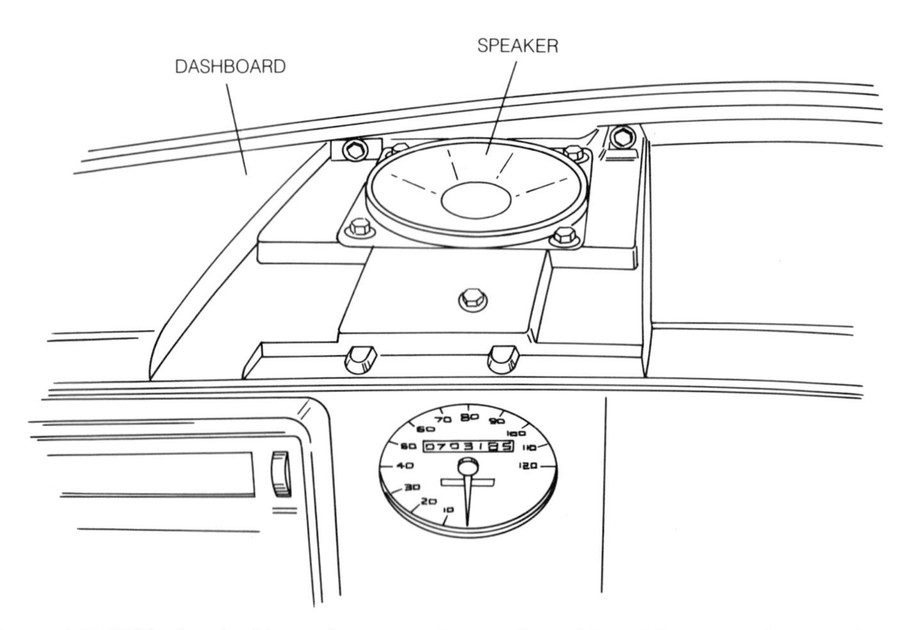

Figure 4-7. With the dashboard removed, you should be able to see the speakers.

MAKING SPEAKER CONNECTIONS

When connecting speaker leads to a speaker, it is essential to make a good solid connection. Most speakers come equipped with flat quick-disconnect terminals with holes in the middle as shown in *Figure 4-8.* These terminals are usually marked with a "+" and "–" or red (+) and black (–) so you can connect them with the correct polarity. The polarity is important so the speakers will be properly phased. Hopefully, your existing harness connectors will match your replacement speaker terminals, and you can use them directly. If not, you can clip the wires and easily solder on new connectors.

With bare wire, strip approximately 3/4″ of insulation from the wire end and stick the bare wire through the hole. Pull the stripped portion through the terminal hole until the insulation almost, but not quite, touches the terminal. While being careful not to break the short flexible wire to the speaker's voice coil, wrap the excess bare wire tightly around the terminal. Use enough heat to make a good solder connection, but be careful not to use excess heat that could loosen the voice coil lead wire.

Of course, you should use good solder connections throughout your system upgrade. A poor connection could become a short circuit that could cause permanent damage to your amplifier or other system component. Also, a poor connection could become a loose, intermittent connection that will cause a "static-like" noise or intermittent loss of the sound when the connection vibrates.

An inexpensive soldering iron may be purchased from a hardware store, and will be well worth the investment. Many hardware stores carry 40-watt pencil tip irons that are ideal for any connections you may make in your projects. Many professional installers use butane-powered soldering equipment because it provides cordless convenience.

PLATE-MOUNT SPEAKERS

As we've discussed, flush-mount speakers are preferred because they can be concealed. However, if space is limited, such as in a compact car or pickup, a plate-mount speaker is a great way to squeeze high-quality sound into snug areas. For example, if the largest flush-mount speaker that will fit into your vehicle is 5 1/4," a 3-way plate-mount, such as the one shown in *Figure 4-9*, can fit into this area because only a small diameter hole is necessary for the woofer and it offers a surface-mount mid-range and tweeter. This type of installation offers a sound similar to a 6" x 9" flush-mount with minimal depth and hole clearance.

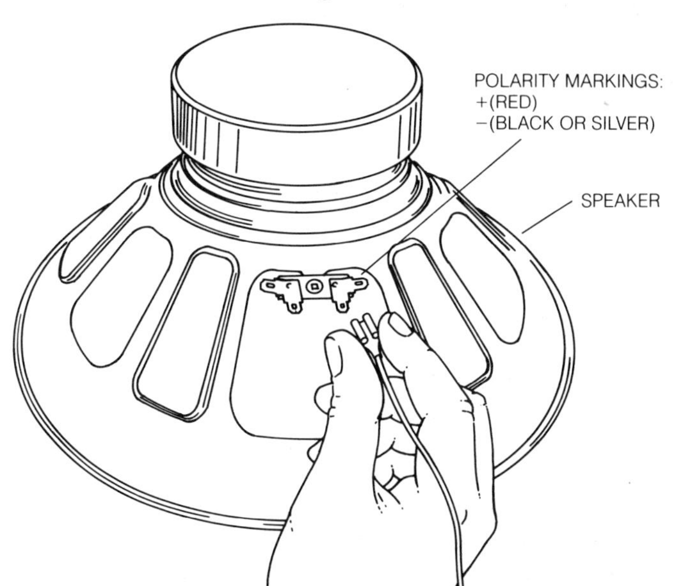

Figure 4-8. Most speakers have flat quick-disconnect connectors with a hole in the middle.

SURFACE-MOUNT SPEAKERS

Another popular alternative to add enhanced sound without cutting speaker holes is the surface-mount speaker. This type of speaker is a good choice for installation in company vehicles and lease vehicles where cutting holes is not appropriate. Because surface-mount speakers are usually fully contained with their own woofer, midrange and tweeter installed in a base reflex design box, their installation requires only two to four screws to hold them in place. Once secured in place, proper polarity hookup of the two speakers wires is all you have to do before you enjoy good sound.

Figure 4-9. A 3-way plate mount assembly that flush mounts into a space normally used by a 5-1/4" speaker. *(Courtesy of Radio Shack)*

CONSTRUCT YOUR OWN SPEAKERS

Instead of purchasing a prefabricated surface mount speaker, you may decide to construct your own. An example of an enclosure built to fit a 6″ × 9″ speaker is shown in *Figure 4-10.* To construct this box, you need a handsaw, wood glue, general-purpose, hardened steel screws, drill, saber saw, tape measure, small wood file, sandpaper, and your choice of wood (either "void free" A-B grade plywood or particle board at least 5/8″ thick).

Before you make the box piece layouts, measure the depth of the speaker you have chosen and assure yourself that you have enough depth in the enclosure. If you don't, increase the depth dimensions or the height of the back to accommodate your speaker. Variations of 10-20% should not affect performance.

After you've cut the six sides of the box to the dimensions shown in *Figure 4-11,* you are ready to cut the actual speaker opening using the template that comes with the speaker. An easy way to start the cut is to drill a hole just inside the cut line of the intended speaker hole as shown in *Figure 4-12.* Be sure to leave enough wood around the four screw holes for secure anchoring of the speaker. When you have finished cutting the hole, try the actual speaker to be installed to see if it fits securely. Use a wood file to smooth the edges so that the speaker will fit flush against the wood.

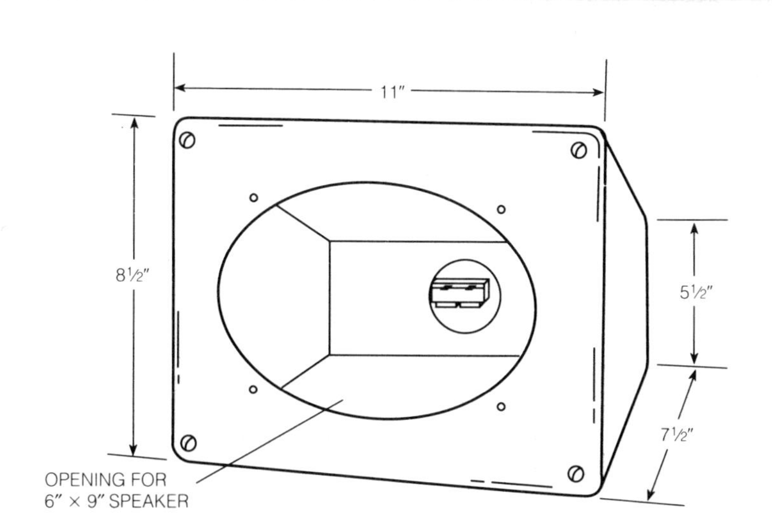

Figure 4-10. Basic enclosure for a 6″ × 9″ speaker.

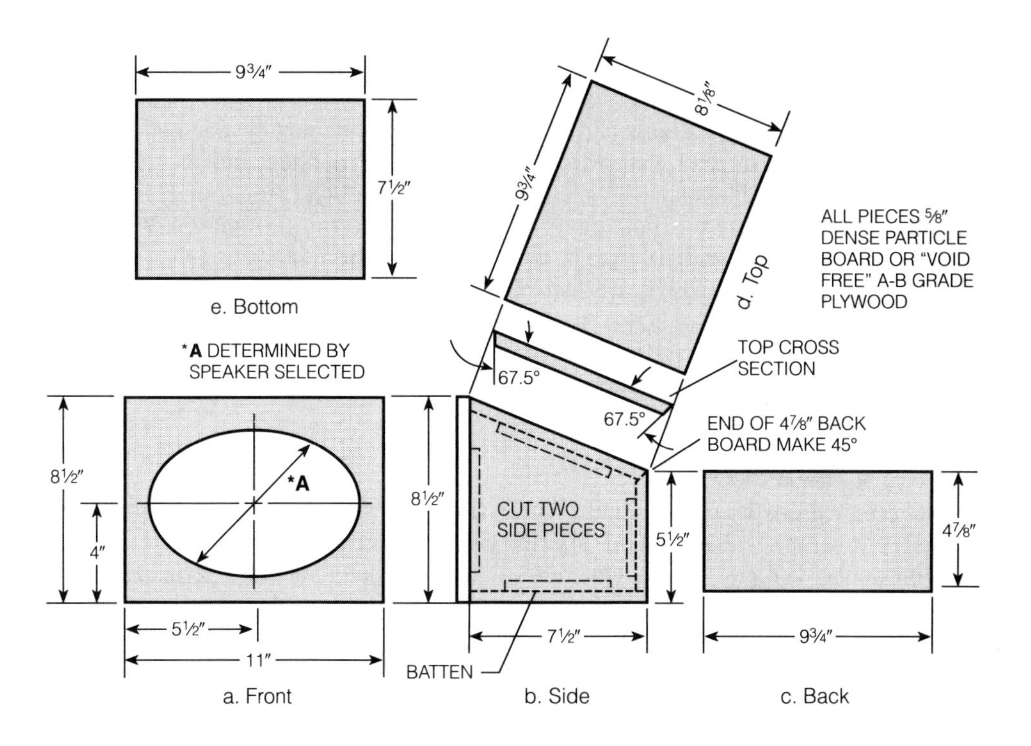

Figure 4-11. Cutting layout for 6″ × 9″ enclosure.

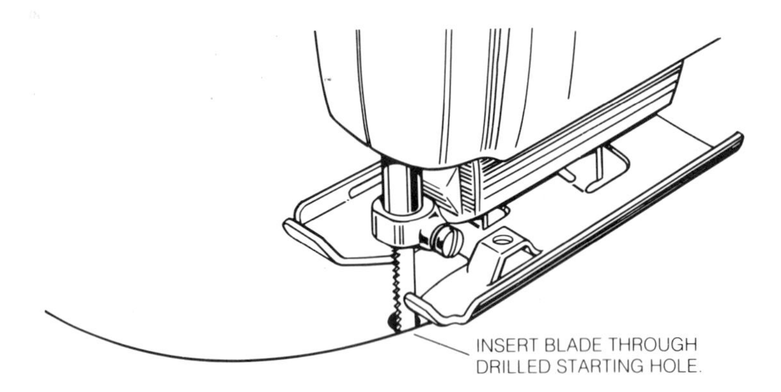

Figure 4-12. Speaker opening cutouts can be made with a power saber saw. Drill a hole inside the cutout line as a starting point for the blade.

Drill a small hole in the back of the box for the speaker leads to go through, or you can cut a hole to use a terminal plate. Before you assemble your box, do a trial fit and use the wood file and sandpaper to make any adjustments. Predrill clearance and pilot holes for screws. When everything fits properly, apply a thin bead of wood glue to the mating surfaces, and screw them together securely to provide an airtight structure and prevent annoying vibrations.

You will probably want to dress up the appearance of your homemade box by applying some kind of finish. A simple way is to apply two coats of a good quality paint or use carpet mastic to attach carpet scraps. When the box is finished to your satisfaction, run the speaker leads through the hole, then connect and solder them to the speaker and terminal plate. Install the speaker in the box using four 1″ hardened steel screws. Attach the terminal plate. If you didn't use the terminal plate, leave a little slack in the wires, and put glue in the hole to hold them in place.

Small "L" brackets, which are available at your local hardware store, may be installed on the sides of the box to secure it to the vehicle's interior. Don't forget to install the grilles to protect the speaker from damage. You can easily remove surface-mount speakers for use in other automobiles, but remember that they are visible and thieves can remove them just as easily.

Adding a Subwoofer

At this point, there is a very good chance that you will want to add a subwoofer enclosure to your speaker system. Specific construction plans and tips pertaining to your particular vehicle will be covered in subsequent chapters. See in particular Chapter 5 for subwoofer electrical connections and filling enclosure with damping material.

SUMMARY

In conclusion, the speakers you choose as upgrades are the backbone of your system. Bargain shopping and taking shortcuts in the installation will prove to be detrimental to the overall outcome of your improvements. Following the advice in this chapter with patience and determination will result in your speaker replacements enhancing your automobile stereo system.

COUPE/SEDAN SYSTEMS 5

This chapter focuses on the installation of stereo modifications in a coupe or sedan-style car. The coupe has two doors and a trunk; the sedan has four doors and a trunk. Although there are hundreds of different makes and models of coupes and sedans, they share many similarities. You should be able to apply the generalized information in this chapter to your specific installation.

NOTE

Before beginning your installation, please read Chapters 1 through 4 completely. There are many mounting and wiring details given that are required in the installation.

SYSTEM PERFORMANCE GOALS

As mentioned in Chapter 3, planning your system is very important; therefore, the first goal in your stereo project should be to establish what you expect from the completed system. Our personal choices would include:

1) AM/FM stereo head-unit
2) Passive equalizer for precise tonal adjustment
3) Compact disc player for its superior sound reproduction
4) Non-distorted, adequately loud sound
5) Good, tight, low bass
6) Crisp, accurate, high-end sound
7) Components neatly installed out of the automobile passenger's eyesight
8) Speakers which are practically invisible
9) Adequate alarm to protect the system

These are the performance features that we deem important. You may have different ideas for your system. Now is the time, before you make any purchases or do any installation, to take a few minutes and make a priority list of your goals for your system.

Don't scrimp on the components necessary to attain the quality sound that you want as a final result. To achieve your goal, you may have to do your modifications in stages because of financial limitations. Therefore, when planning your system, choose components that can function alone, but also can be used with other components as you expand your system.

Start with a good head-unit and a pair of good speakers. The head-unit should be capable of pre-amp (line-level inputs and outputs) operation. This will allow you to add a passive equalizer for tonal adjustment and separation, and power amplifiers for increasing the power to the speakers — especially for driving a subwoofer for the low-end sounds. Don't economize on your speakers; if they are inadequate, none of the other components can deliver up to their potential. You can add a component or two at a time until you obtain the system you planned.

SYSTEM REPLACEMENT

Basically, only two styles of in-dash head-units are available on the market today: (1) the traditional double-shaft style now commonly called DIN-C; and (2) the newer DIN-E "flat-faced" design found in most late-model automobiles. DIN-E requires only a single rectangular opening.

If you decide to purchase the DIN-C double-shaft style, here are a few things to take into consideration. First, remove the existing radio. If the automobile you are working on was made in 1979 or before, then there is a good chance it is set up for a double-shaft style radio. This was the most popular style of radio for pre-1980 cars. If the hole left in the dash after disassembling the factory head-unit looks similar to the one shown in *Figure 5-1,* then much of the work may be completed already. You may not have to change the openings. Take all of the measurements shown in the figure: width, height, shaft hole diameter, and distance between the centers of the two shaft

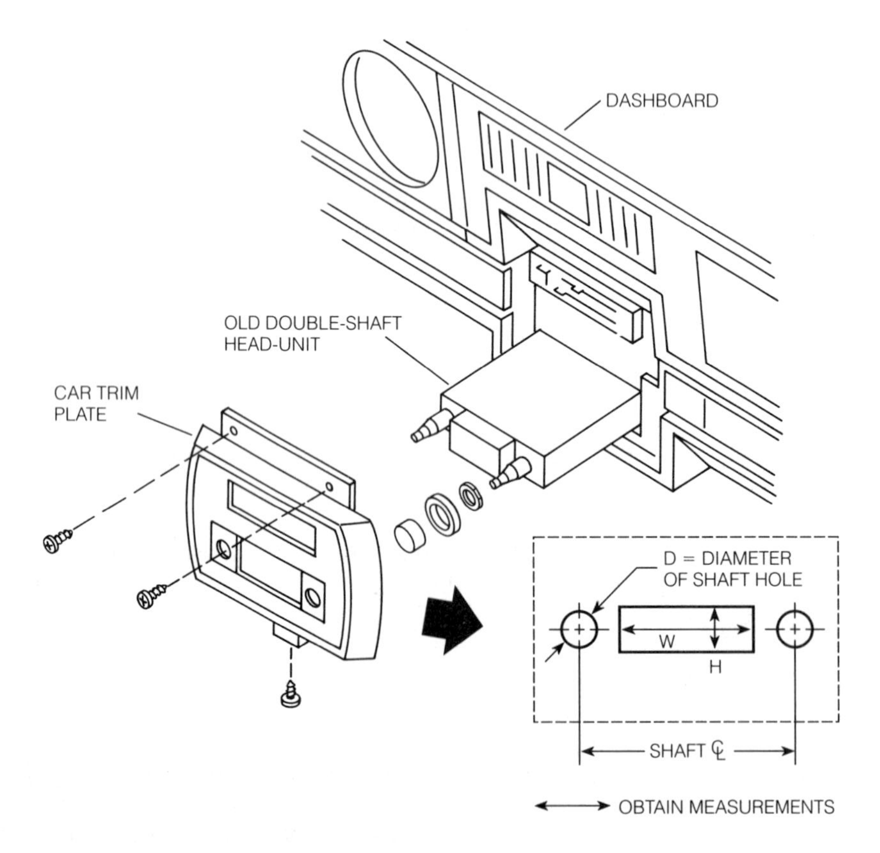

Figure 5-1. After removing the old double-shaft head-unit, obtain the indicated measurements to help you select the proper replacement unit.

holes. Also, measure the available space behind where the radio mounts for adequate mounting depth. This usually is not a problem, but some vehicles have very limited space. Take all of this information with you when you shop for your head-unit to help you choose a unit that will fit neatly into the automobile without modifying the mounting holes.

Example Installation

For our example in this chapter, we removed the factory system from a 1986 Oldsmobile Cutlass® Supreme Classic Coupe and replaced it with a system of after-market upgrades consisting of:

Item
DIN-C (double shaft) AM/FM stereo with cassette player
7-band passive stereo equalizer with subwoofer crossover
Two 80-watt (40W per channel) stereo power amplifiers
One 160-watt (80W per channel) stereo power amplifier
Pair of dynamic horn tweeters
Pair of 4-way surface-mount speakers
Two 8″ subwoofers with dual voice-coils (for enclosure)
Pair of speaker grilles
In-dash head-unit installation kit (1982-90 GM)
Two terminal plates for speakers
Material for enclosure
Wire, connectors, cables, hardware, etc.

HEAD-UNIT

We chose the head-unit shown in *Figure 5-2* because of these three features:
- Line-level outputs which will be used to connect to the passive equalizer (which, in turn, will drive the power amplifiers)
- Above average FM tuner features
- Superior auto-reverse cassette player functions.

It required an installation kit.

CAUTION

Before attempting any installation, you should disconnect the negative terminal connection from the automobile battery to prevent damage to the vehicle wiring or the equipment you are installing.

When you have purchased your system and you are ready to make your installation, carefully read all of the manufacturer's instruction included with the head-unit. If a mounting kit is required, and for our installation we used the installation kit shown in *Figure 5-3*, thoroughly read the instructions for its use. Carefully follow the step-by-step instructions to install the head-unit securely. See Chapter 3 for the wiring hookup procedure for the signal and constant power for preset memory, ignition-switched power, display and ground connections. Many electronics stores have interfacing cables and connectors if you need them.

Figure 5-2. This is an in-dash-mounting DIN-C, double-shaft, AM/FM head-unit. *(Courtesy of Sanyo)*

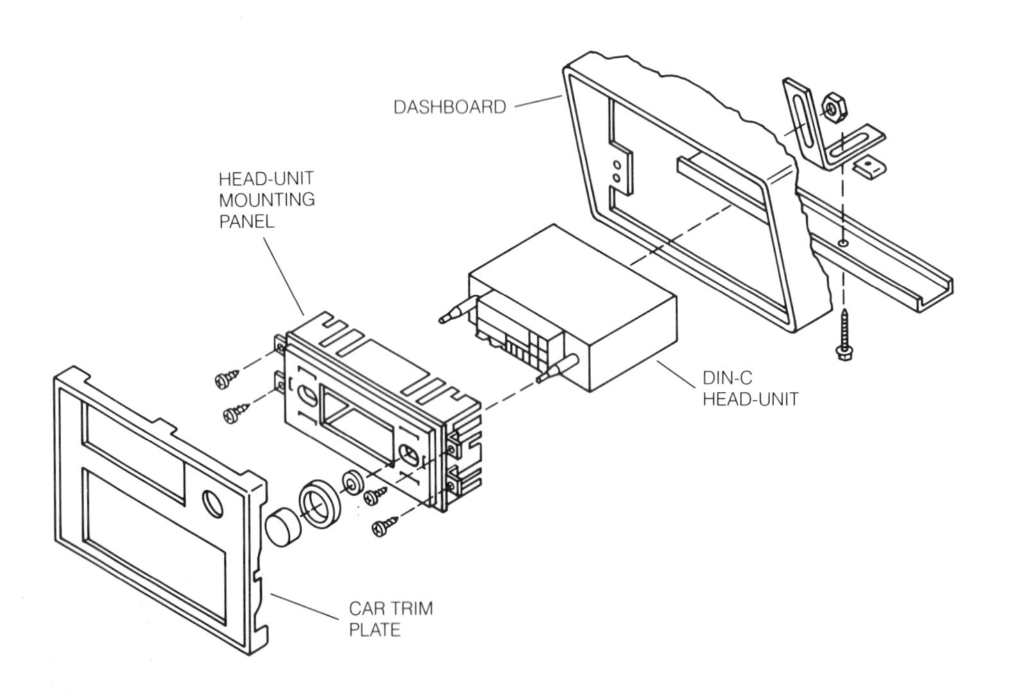

Figure 5-3. This installation kit permits in-dash mounting of a DIN-C head-unit with two shaft openings and a rectangular opening.

NOTE

The installation kit of Figure 5-3 is for a DIN-C style head-unit which has a rectangular opening and two separate shaft openings. For a DIN-E head-unit, the single rectangular opening must be cut out from the kit's head-unit mounting panel face on lines already scribed into the molding. In some kits, a special adapter plate is available. Chapters 6 through 9 have more details to help you make the DIN-E choice.

EQUALIZER

An equalizer, if your plan includes one, is the next step after the head-unit is in place. For our installation, we chose one that has a number of our favorite configurations. Its wide-range passive tonal adjustment and built-in crossover for the subwoofer output permits continuous adjustment of the sound over a large dynamic range and wide frequency spectrum. Also, it has a slim-line outline and can be installed in either of two ways.

We used the first way of installing one by securing it with a bracket under the dash as was shown in *Figure 3-5*. A second way is to use an in-dash installation kit that accepts both a head-unit and a slim-line equalizer. Installing a slim-line equalizer in the dash with such a kit provides a very neat, custom look. Your installation must have the depth in the dash area to mount both the equalizer and the radio.

An installation kit similar to the one needed, except it is for a DIN-E flat-faced head-unit, is shown in *Figure 6-3*. The correct kit contains an adapter plate for the DIN-C head-unit and the equalizer. These kits, made by several manufacturers, are available from outlets that install automobile audio equipment. Precisely follow the instructions that come with the installation kit to mount the equalizer. Remember, these kits are designed for use only with a slim-line equalizer like the one we chose. After mounting is complete, refer to Chapter 3 for the proper wiring hookup.

POWER AMPLIFIER

If your plan includes more power, an amplifier or a power amplifier system is your next step. For our example installation, we decided to include two 80-watt (40W per channel) stereo amplifiers and one 160-watt (80W per channel) stereo amplifier. Both of these amplifiers have isolated inputs to reduce electrical noise pickup; this is an important factor in improving the overall signal-to-noise ratio of the total system. Also, both of these amplifiers have line-level input jacks which will match the outputs of the stereo equalizer, including the equalizer's subwoofer output jack. In addition, these amplifiers are fully overload protected in case of an accidental short circuit in the speaker wiring.

Mounting the amplifiers is a relatively simple task. Preferably, an amplifier should be out of the passengers' immediate eyesight. It can be mounted under a seat, behind a seat, in the trunk, or anywhere it can get adequate ventilation for cooling. Most amplifiers come with a set of mounting brackets and mounting hardware. Usually, by using self-tapping screws, you can firmly secure the amplifier to the automobile's chassis as shown in *Figure 5-4*. After securely mounting the amplifier or amplifiers, refer to Chapter 3 for general wiring instructions. For our specific example, the

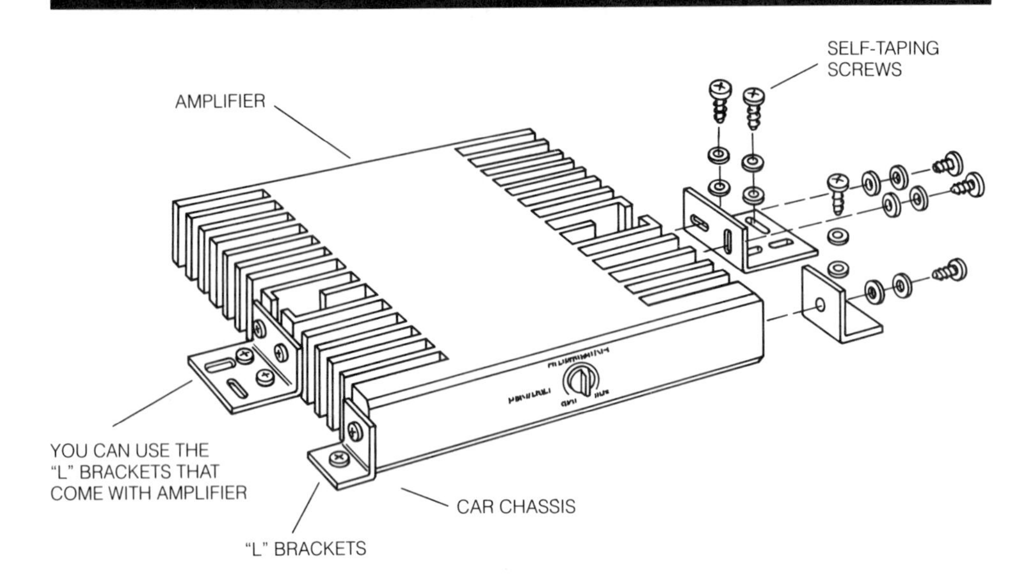

Figure 5-4. This is a good way to secure an amplifier to a car chassis.

complete system wiring is shown in *Figure 5-12.* We chose to mount an amplifier under the front seat to drive two tweeters installed in front factory openings. The amplifiers to power the rear deck speakers and the subwoofers were mounted in the trunk.

CAUTION

If the amplifier is not securely mounted, it will probably move around. Stress on the wires and components due to this motion may cause wire breakage and/or permanent damage to the amplifier. A loose amplifier could become a dangerous missile when the vehicle comes to a sudden stop.

CHOOSING THE SPEAKER SYSTEM

In a coupe/sedan style automobile, we recommend that you first identify factory speaker openings as we discussed in Chapter 4. Often, they can be found in the dash, doors and rear deck. If no subwoofer is in your plan, then full-range, high-quality speaker replacements are recommended in at least four of the factory openings, e.g., dash/doors, dash/rear deck. In almost any installation, four speakers will provide a relatively full and enhanced listening area; whereas, only two speakers give a one-sided, single-direction sound, often referred to as "tubular." These speakers can be hooked to the audio output of the self-contained head-unit or to external amplifiers, whichever you choose.

In our specific example, we chose to install new tweeters in the in-dash factory holes and a set of surface-mount enclosures on the rear deck. Each of the surface-mount enclosures contains a woofer, mid-range speaker and dual tweeters. The surface-mount enclosure was securely mounted to the rear deck with nuts and bolts. We drilled holes through the rear deck to run the speaker wires to the amplifier. (If factory speakers are already mounted in the rear deck, you can remove them and place the surface-mount enclosure directly over the hole, or for a neater installation, place a separate fabric-covered board between the surface-mount enclosure and the rear deck.)

CAUTION

As discussed in Chapter 2, if you choose external amplifiers, do not overpower your speakers.

Use of a Subwoofer

There are several choices if you are going to use a subwoofer. One choice is a special woofer that has a stiffer suspension and is designed to operate without damage in free-air without an enclosure. It utilizes the trunk's air space as its enclosure. Another choice is a small prefabricated box or your own custom-built subwoofer box, which, in either case, will be positioned in the trunk behind the rear seat. Building the subwoofer box yourself can boost your pride and feeling of achievement that you will have for your completed system, but remember that the size of the enclosure is sometimes limited in a coupe/sedan style automobile.

If you do not want to build a speaker enclosure and have found a special free-air style woofer, then place the woofer on the rear deck behind the rear seat. Be sure to choose a speaker that is not too large to fit in the rear deck area. Most rear decks will accommodate a woofer of 6½″ — some up to 8″ — in diameter. Measure the length and width of the deck space and the diameter of the speaker to determine if it will fit in the proposed area.

Remember, if you use the rear deck for woofer placement, there likely will be no additional room for a full-range speaker to handle the mid-range and high frequencies. In a four-door sedan, this does not present a problem because full-range speakers may be placed in the factory speaker holes that are often located in the rear doors. In a coupe, the front door factory speaker openings can be used for full-range speakers. We do not recommend full-range speakers in the dash and woofers in the rear deck because of an overpowering bass sound, poor "surround-sound," and a very limited listening area.

Locating Rear-Deck Opening

The speaker holes required in the rear deck can be easily cut using a power saber saw equipped with a metal cutting blade. Most speakers come with a template. Lay the template in the exact area to be cut and trace the perimeter for cutting.

CAUTION

Carefully examine underneath the rear deck area to make sure that you will not be cutting into wires or critical structural members in the trunk.

The position of the mounting hole may be dictated by the clearance that is available for using a saber saw to cut the opening. If your rear glass is set at an angle so that you cannot use a saber saw on top of the rear deck, you may have to cut from the bottom of the deck. However, cutting from the bottom can be much more difficult than from the top because of the unusual structural members in the trunk.

Always wear safety glasses when drilling, sawing and filing to prevent eye injury from flying particles.

If no template is included, you will need to accurately measure the speaker's diameter. *Figure 5-5* shows the appropriate way to measure the diameter for the speaker opening. Make a pattern to trace or follow for cutting.

CAUTION

You can always cut a little more if the hole is too small, but it is practically impossible to replace material if you cut the hole too big. Be careful!

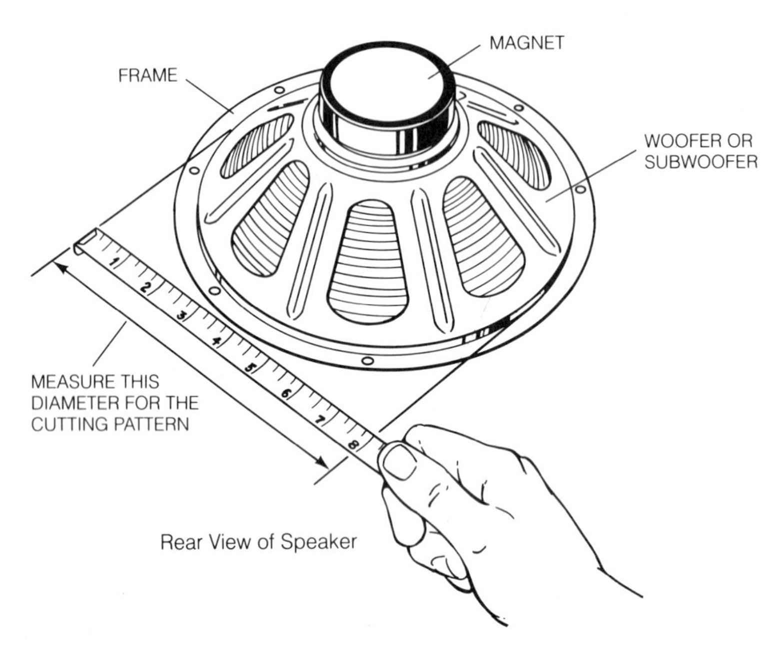

Figure 5-5. This is a way to obtain the diameter for cutting a speaker opening in an enclosure.

Once you have marked the speaker hole, you are ready to cut. Drill a small hole inside the cutout and start cutting as shown in *Figure 4-12.* Usually you are cutting through cloth, fiberboard, plastic, and metal. After you have finished the cut, set the speaker in place. If the speaker does not fit perfectly, remove the speaker and use a coarse file to modify the speaker opening.

Once the speaker fits snug and flush, you are ready to anchor it to the deck. The speaker must be anchored securely to prevent annoying rattle and to help achieve an airtight fitting. We recommend using self-drilling screws in all the anchor holes around the edge of the speaker (hint: if you angle the screws slightly away from the speaker it will help ensure a firmer grip). Be careful, a slip of the screwdriver could puncture the speaker cone. After you have completed the mounting, you can hook up the speaker leads. Use a speaker grille to protect the speaker from damage. Several types are available from most radio dealers.

Wiring a Dual Voice Coil Woofer

Subwoofers like the 8" ones chosen for this installation and the 12" ones for other installations discussed in later chapters have dual voice coils. Each voice coil has

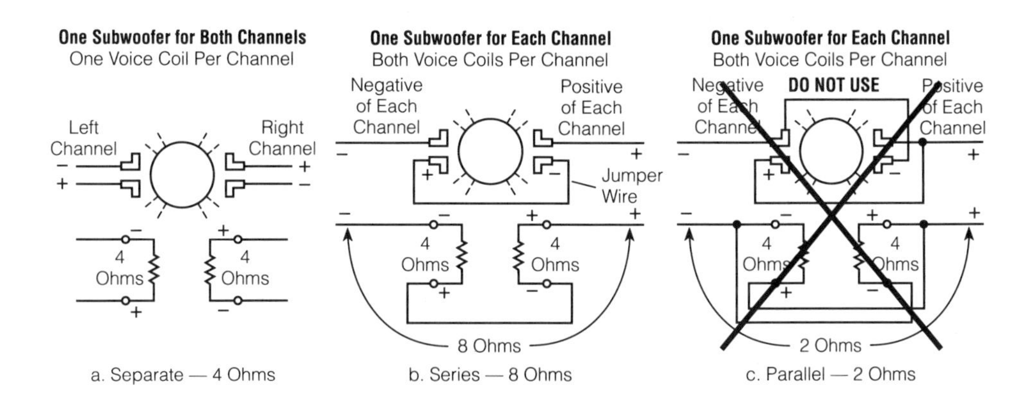

Figure 5-6. Dual Voice Coil subwoofer connections.

its own two terminals as shown in *Figure 5-6.* Each voice coil has 4 ohms of resistance. Three possible connections are shown. *Figures 5-6a* and *5-6b* can be used. *Figure 5-6c* SHOULD NOT BE USED.

The connection of *Figure 5-6a* is used when one subwoofer is used for both channels of a stereo amplifier (see *Figure 9-2*). Connect the right channel + and – outputs of the amplifier to one set of + and – terminals on the subwoofer as shown in *Figure 5-6a.* Connect the left channel + and – outputs of the amplifier to the other set of + and – terminals on the subwoofer. Each amplifier channel output must drive 4 ohms.

The connection of *Figure 5-6b* is used when a subwoofer is used for *each* channel of a stereo amplifier or the output of a monaural amplifier. As shown in *Figure 5-6b,* connect a jumper wire from the + terminal of one set of + and – terminals of the subwoofer to the – terminal of the other set of + and – terminals. Connect the channel (either left or right) output + and – terminals to the remaining + and – terminals on the subwoofer. The dual voice coils in *Figure 5-6b* are connected in series so the amplifier channel output is driving 8 ohms.

The connection of *Figure 5-6c* SHOULD NOT BE USED. It has the dual voice coils connected in parallel and each amplifier channel output would have to drive 2 ohms. The amplifier output would be required to supply excessive current to supply rated power.

Building Your Own Enclosure

If you decided to build your own enclosure, you will need some basic shop tools, including:

- Circular saw for the large square cuts
- Power saber saw for cutting circles
- Standard claw hammer
- Minimum 10′ measuring tape
- Carpenter's square
- Assorted wood rasps, files and sandpaper
- Drawing compass to draw speaker opening cut-outs
- Finishing nails
- Hardened steel screws (#8 × 1″)
- Punch to set nails
- Drill and drill bits
- Staple gun (not an office stapler)
- Carpenter's glue
- Sealing caulk (white silicone)

If you do not own a power saber saw, consider renting or borrowing one for the duration of this project. If you do not have a circular saw, check with places that sell lumber. Often they will cut pieces to a customer's specifications for a relatively small fee. This can help ensure square cuts and provides smaller pieces for easier transportation to your home.

Assembling the Box

Once you have gathered the necessary equipment, you are ready to construct your box. Inside the trunk, measure the height from the floor of the trunk to the bottom of the rear deck; then measure between the inside edges of the trunk lid opening. This will give you a general idea of the maximum size box that will fit in this area.

Figure 5-7a shows the dimensions of the basic trunk-mounted box with two 8″ woofers that we used in the Oldsmobile Cutlass. Compare these dimensions to your available trunk space. If it will fit, we recommend using these dimensions. Allow room for the trunk lid hinges to come down without hitting the box. The dimensions can be modified to meet your particular needs. A variation of 10% will not noticeably affect the sound characteristics. An additional 20% increase in effective volume can be obtained by filling the enclosure with DACRON® polyester fiberfill available at department stores and sewing centers. Place netting over the back of the speaker and staple it in place so fiberfill doesn't contact the speaker cone.

When you have the dimensions of your enclosure, purchase the wood necessary for the project. Cut (or have cut) the pieces of the enclosure. Use the cutout patterns shown in *Figure 5-7b.* The front has the speaker openings cut into it as shown in *Figure 5-7b.* The "D" dimension is adjusted for the particular speaker chosen. Measuring and marking for the speaker openings are done as we discussed earlier. Do not cut the holes too large; you can always enlarge them a little more with a wood rasp.

Next, construct the box. You can start with any two adjacent sides, or one of the sides and the bottom. First, assemble all parts together without gluing. Drill clearance and pilot holes for screws, and make sure everything fits well. When ready for final assembly, use a good quality wood glue — yellow aliphatic resin is recommended — and apply glue along one edge of the wood, then secure the panels together as shown in *Figure 5-8.* The battens or blocks are used to strengthen the enclosure's joints. Assemble the front, back and sides to the bottom. Lay the top aside to be assembled later. Seal all joints with caulk as shown in *Figure 5-9.* You may mount the speakers on the inside or outside of the front speaker board. We chose to mount them on the outside after the exterior of the box was finished.

Next, we recommend that you cut openings for two quick-connect terminal plates like those shown in *Figure 5-10.* When installed on the box, they will provide easy connection of speaker leads. Now, assemble the top to the battens that were previously fastened to the sides. Caulk all top joints and use the rasp and sandpaper to smooth out the rough edges of any cuts and holes.

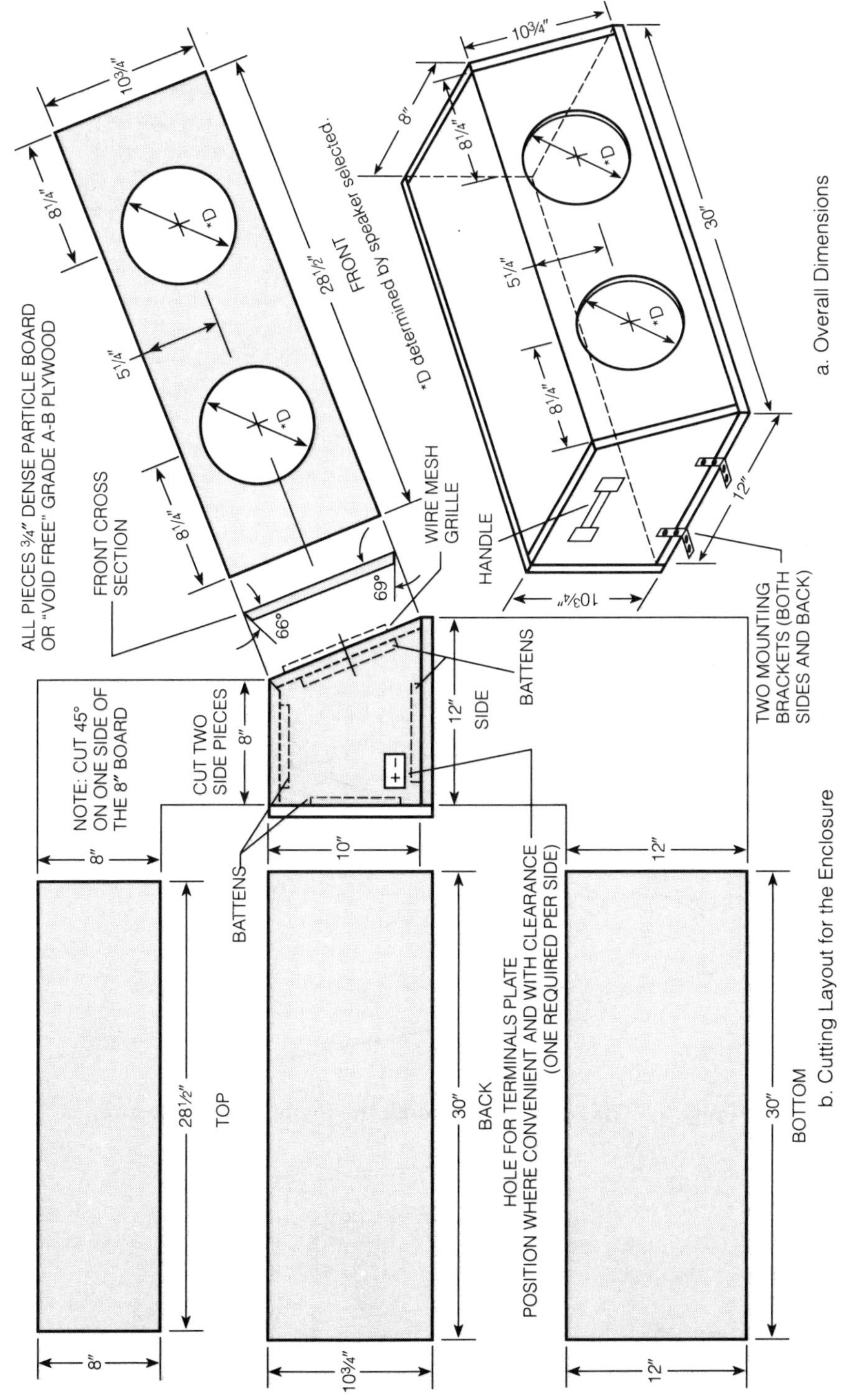

Figure 5-7. Cutting layout for trunk-mounted enclosure.

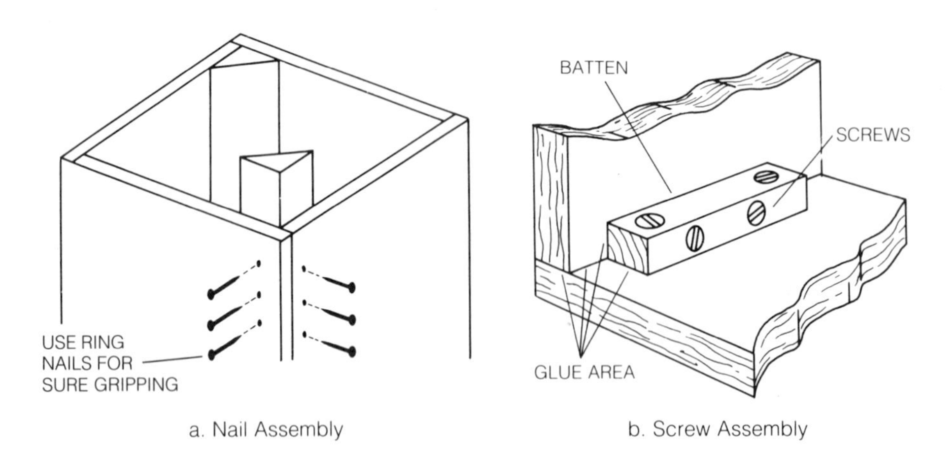

Figure 5-8. This is a good way to fasten corners of an enclosure.

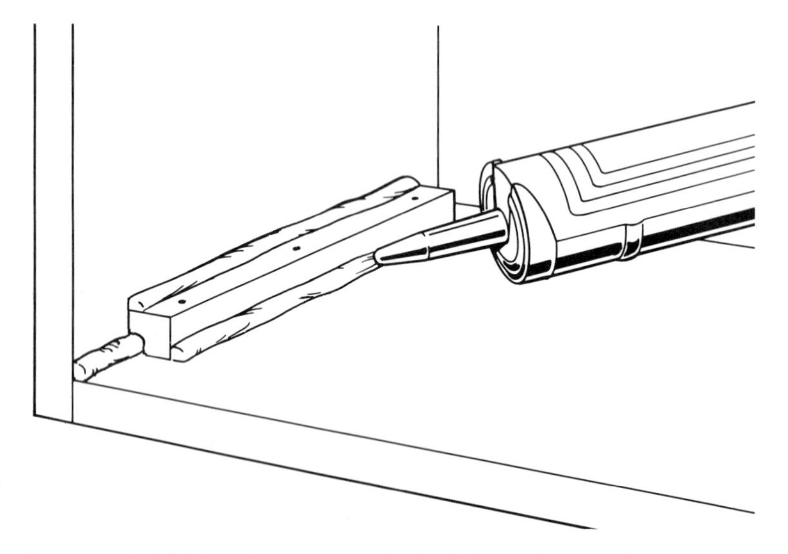

Figure 5-9. This is a way to caulk the joints of an enclosure.

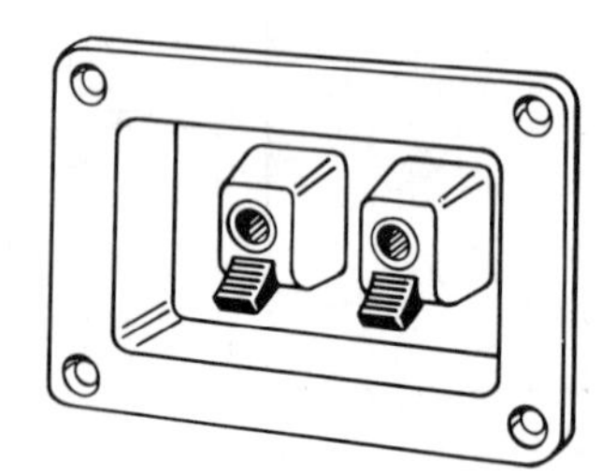

Figure 5-10. Quick-connect recessed-mount terminals for a speaker enclosure.

Finishing the Box

A nice finishing touch is the application of carpet to the enclosure's exterior to make it more appealing to the eye. Various types and colors of carpet are available at most carpet stores. An exact match to the car's factory installed carpet can usually be found at the car dealer's parts department. To apply the carpet, cut pieces to fit the exterior walls of the box, then use carpet mastic cement or a staple gun to affix it properly. Once the carpet is in place, use a single-edge razor blade or a utility knife to trim it and to cut the openings for the speaker and quick connect pads.

Solder wires to the inside terminals of the quick-connect terminals. Make them long enough to reach through the box and outside the speaker openings. Solder the other ends to the speaker terminals. Since a subwoofer is being used for each channel, the series connection of *Figure 5-6b* is used. Be sure to keep polarity correct. Install the quick-connect terminal assemble and secure the speakers to the box with 1" hardened steel screws. Finally, install wire-mesh speaker grilles to protect the speakers.

Anchoring the Enclosure

Now that the box is complete, you are ready to secure it in place. We recommend that you use small "L" brackets mounted to the bottom edge of the box as shown in *Figure 5-7.* Set the box in place in the trunk. Make sure it is pushed up snugly against the rear seat. Use self drilling screws to anchor the box in place.

Once anchored in place, connect the wires from the amplifier to the quick connect terminals and the job is complete. Again, be careful to observe polarity when connecting any of the speaker wires. The final installation should look similar to the one shown in *Figure 5-11.*

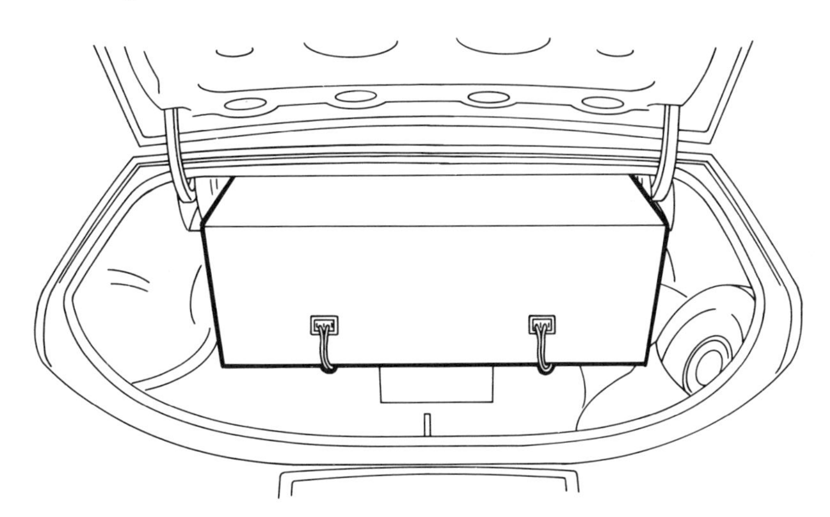

Figure 5-11. Completed box enclosure installed in the trunk of a 1986 Cutlass Supreme Classic coupe.

We recommend that you perform a final check of all wire connections before reconnecting the vehicle battery terminal. The complete final wiring system is shown in *Figure 5-12.* Refer to *Figures 3-6* and *3-7* if any special cables or connectors are required. The wires from the equalizer to the amplifier(s) may be run under the carpet. If the amplifiers are mounted in the trunk, then their wires to the speakers can be connected directly. Once you are satisfied that the wiring is complete and correct, reconnect the vehicle battery terminal and enjoy your new system.

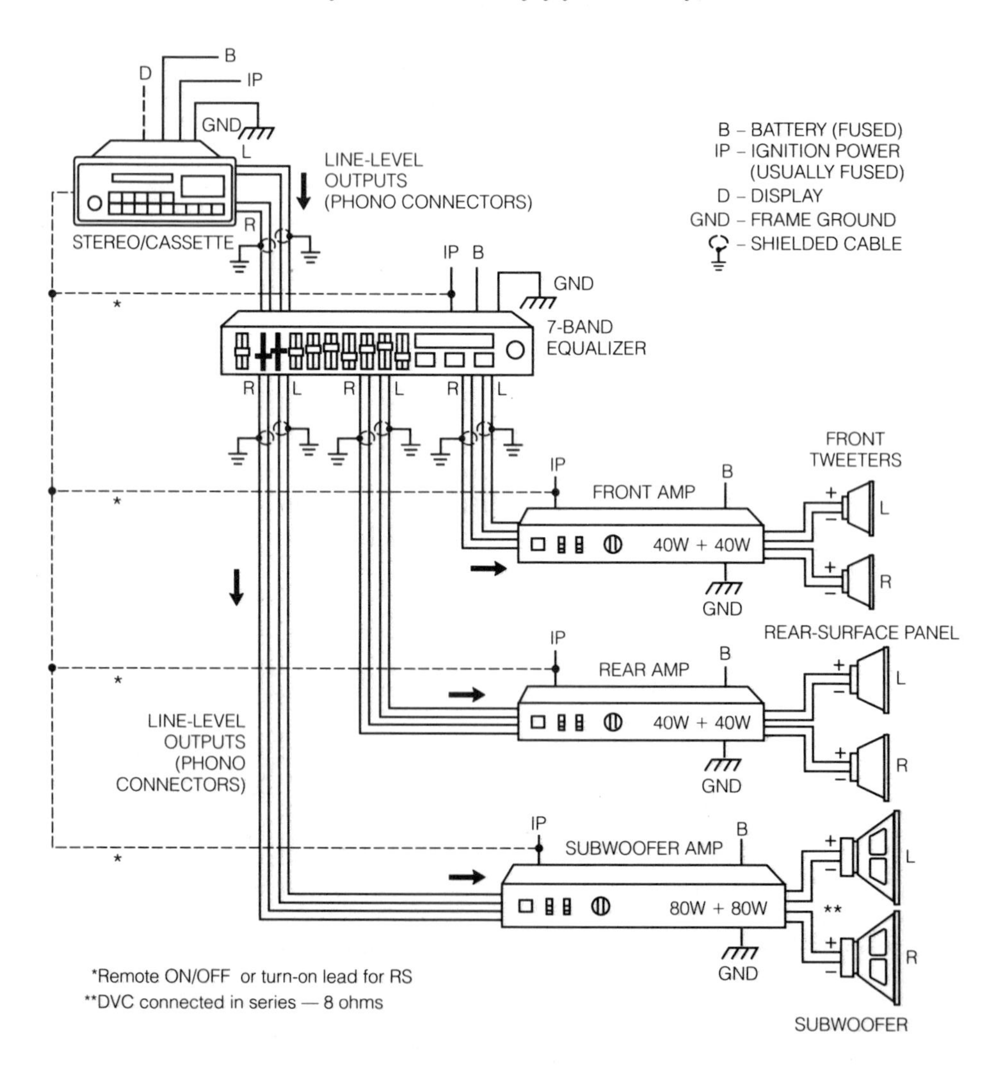

Figure 5-12. Wiring diagram for the sound system installed in a 1986 Oldsmobile Cutlass Supreme Classic coupe.

HATCHBACK SYSTEMS 6

HATCHBACK SYSTEMS

This chapter focuses on installation of stereo modifications in a hatchback-style automobile. The hatchback is a two-door automobile with a rear lift gate. It is similar in many ways to a coupe or sedan. The difference is that there is no trunk. For subwoofer sound in the hatchback, a fabricated speaker enclosure is necessary. We will discuss this kind of installation as one example in this chapter.

Just as with the coupe and sedan, there are many different makes and models of hatchbacks. It is, therefore, impossible to give step-by-step instructions for each make and model. However, these models share many similarities, so you should be able to apply the generalized information presented here to your specific hatchback system installation.

We have chosen to describe the design and installation of two complete systems in this chapter. The first is a high performance system for a 1989 Pontiac Firebird and the second is an good quality economy system for a 1987 Toyota Corolla.®

NOTE

Before beginning your installation, please read Chapters 1 through 4 completely. There are many mounting and wiring details given that are required in the installation.

SYSTEM PERFORMANCE GOALS

Once again, planning is very important so the first goal in your stereo project should be to establish what you expect from the completed system. In Chapter 5, we listed some of the personal choices that we would include in our favorite system. You might use that list as a starting point for making a list of characteristics for your system. You may have some different ideas, or want different components for your system. In either case, it is important that you take time to plan and make your priority list before you make any purchases or do any installation.

As we also said in Chapter 5, you will be happier with the completed job if you do not scrimp on the components necessary to attain the quality sound that you want as a final result. To achieve your goal, you may have to do your modifications in stages because of financial limitations. Therefore, as in Chapter 5, when planning your system, choose components that can function alone, but also can be used with other components as you expand your system.

As we recommend for any system, start with a good head-unit and a pair of good speakers. The head-unit should be capable of pre-amp (line-level inputs and outputs) operation. This will allow you to add a passive equalizer for tonal adjustment and separation and power amplifiers for increasing the power to the speakers — especially for driving a subwoofer for the low-end sounds. Don't economize on your speakers; if they are inadequate, none of the other components can deliver up to their potential. Remember that the "weakest link" parable applies to this situation also. You can add a component or two at a time until you obtain the system you planned.

REPLACEMENT COMPONENTS

In just the past few years, a remarkable number of new products have elevated the quality and widened the variety of audio entertainment possibilities. More and more people are installing sound systems in their automobiles that rival or, many times, surpass the system in their homes.

The development of the compact disc has led the way to a significant improvement in the overall sound quality and convenience to the music lover. The Audio Home Recording Act of 1991 broke the deadlock over Digital Audio Tape (DAT). In addition, new and exciting products on the horizon, such as digital compact cassettes, promise dynamic and exciting new formats.

A VERY HIGH-QUALITY SYSTEM

For our first example in this chapter, we will describe a top-notch system designed and installed in a 1989 Pontiac Firebird hatchback. The system was planned to meet the following specifications:

- An AM/FM stereo receiver with a digital electronic tuner, memory-scan sampling, seek, and automatic storage of station selections in its memory.
- A high-quality CD player with automatic features for ease of operation while driving.
- A full-range frequency response with a smooth, detailed sound contour from the crisp high tones all the way down to hard-hitting bass.
- A crossover to direct frequencies to the correct driver so the drivers can reproduce the audio spectrum without interference between drivers.
- Enough power from the system to overcome background road noise (wind noise when T-tops are removed), and to maintain the dynamic range of the signal from soft to loud so the music sounds live.
- The total system neatly installed using the limited space available in the hatchback, but with the components out of view of the driver and passenger.

To meet the desired specifications, the following components were selected:

Item
DIN-E AM/FM stereo with CD player
and pull-out chassis
7-band passive stereo equalizer with subwoofer crossover
One 80-watt (40W per channel) stereo power amplifier
Two 160-watt (80W per channel) stereo power amplifiers
Two 12″ subwoofers with dual voice coils (for enclosure)
Pair of 6″ × 9″ 3-way replacement speakers
Pair of speaker grilles
Installation kit for DIN-E head-unit and equalizer
Material for enclosure top
Wire, connectors, cables, hardware, etc.

CAUTION

Before attempting any installation, you should disconnect the negative terminal connection from the automobile battery to prevent damage to the vehicle wiring or the equipment you are installing.

Installing the Head-Unit

In the coupe sound system of Chapter 5, we described the factors involved in the selection and installation of a double-shaft head-unit. For this Pontiac Firebird hatchback project, we decided to install a DIN-E (flat-faced) head-unit with AM/FM stereo and in-dash compact disc player. The DIN-E style head-unit requires a rectangular opening. Many of the late model vehicles manufactured since 1980 have this rectangular opening from the factory. Vehicles made prior to 1980 are usually set up for a double-shaft style. Either way, a DIN-E style head-unit can be installed.

If a square opening exists in your automobile, then check with the dealer where you are buying your head-unit to determine if an installation kit is necessary. For example, the radio mounting panel in the kit used in Chapter 5 can be cut out so a DIN-E flat-faced unit will fit. If you use a kit, closely follow the instructions included with the head-unit and installation kit.

If you are installing a DIN-E head-unit into a double-shaft slot of an older automobile, then the opening must be cut to the appropriate size as shown in *Figure 6-1*. A template with the precise measurements of the head-unit is often included with a DIN-E radio. Carefully mark the area to be cut and use a sabre saw or jigsaw with a metal cutting blade.

CAUTION

If you are cutting into the dash, check behind the cutting area for wires or anything else that could be damaged by the saw blade. Remember, a sabre saw or jigsaw blade extends back several inches due to its reciprocating action.

After the hole is cut, follow the instructions to install the head-unit securely. You may want to refer to Chapter 3 again for a more complete wiring guide. You may need a wiring harness adapter.

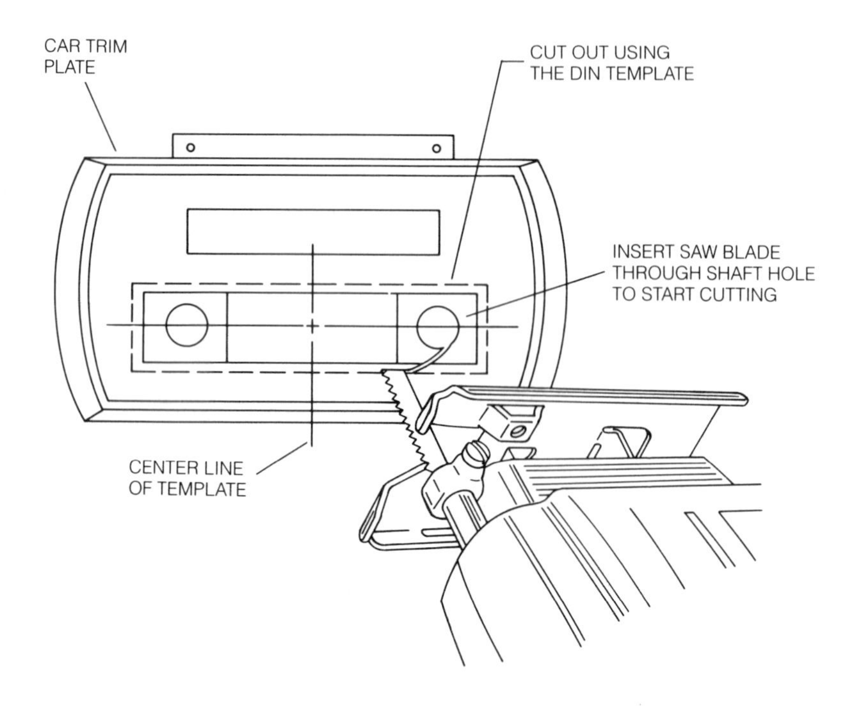

Figure 6-1. Using a template to cut out the double-shaft mounting area to permit mounting a DIN-E head-unit.

Installing a CD Player

For the quality features specified for the system designed into the Firebird, the compact disc player shown in *Figure 6-2* will be installed because of its high performance audio reproduction capability. All in-dash CD players on the market today require a DIN-E cutout.

It can be installed in your automobile as easily as any DIN-E head-unit with cassette. In our installation, we used an installation kit for both the CD player and equalizer. If yours is different, follow the mounting discussion in Chapters 3 and 5. The CD head-unit chosen offers a theft-deterrent pull-out chassis, high-power 16-watt-per-channel output, built-in fader, and preamp (line-level out) jacks for proper external amplifier hookup. Its built-in CD player offers 4X oversampling, programmable CD selecting, random shuffle play, and preset scan, which plays 10 seconds of each track until the desired tune is found. Its built-in AM/FM stereo tuner features eighteen FM and six AM presets for quick station selection, and built-in balance and fader controls for better tonal control. The wiring of a CD player is almost identical to that of an in-dash AM/FM cassette head-unit. See Chapter 3 for wiring details. Electronics stores have interfacing cables and connectors if you need them.

Figure 6-2. This in-dash AM/FM stereo compact disc player has a pull-out chassis. *(Courtesy of Panasonic)*

Installing the Equalizer

If your plan includes an equalizer, then you need to choose between the installation methods discussed in Chapter 5. For the Firebird system, the same equalizer chosen for the coupe/sedan system is installed, but now an installation kit is selected that accepts both the DIN-E radio and the slim-line equalizer together in one panel. The installation kit is shown in *Figure 6-3.* This kit is made by American International and is sold in outlets that install automobile audio equipment. The space in which the equalizer and the pull-out chassis head-unit is to be inserted must be checked carefully so there is enough depth to install both pieces of equipment. Some tailoring of the kit openings may be required.

If the kit required for your automobile looks similar to the one shown in *Figure 6-3*, then your equalizer probably can be mounted in the dash. Remember, these kits are designed for use only with a slim-line (half-DIN size) equalizer. The instructions for the installation kit must be followed exactly. The wiring diagram is shown in Chapter 3.

Choosing the Speaker System

When dealing with a hatchback style automobile, we recommend that you identify all factory speaker openings as we discussed in Chapter 4. In hatchbacks, they are often located in the dash, door, and/or rear quarterpanel. If you do not plan to add a subwoofer, then we recommend replacing the factory speakers with high-quality full-range speakers in at least four of the openings (dash and rear quarterpanel or door and rear quarterpanel). (We will discuss this in greater detail when we cover the system installation of the Toyota Corolla later in this chapter.) As before, in almost any installation, four speakers will provide a relatively full and enhanced listening area; whereas, only two speakers give a one-sided, single-direction sound, often referred to as "tubular." These speakers can be connected to the audio output of the self-contained head-unit or to external amplifiers, whichever you choose.

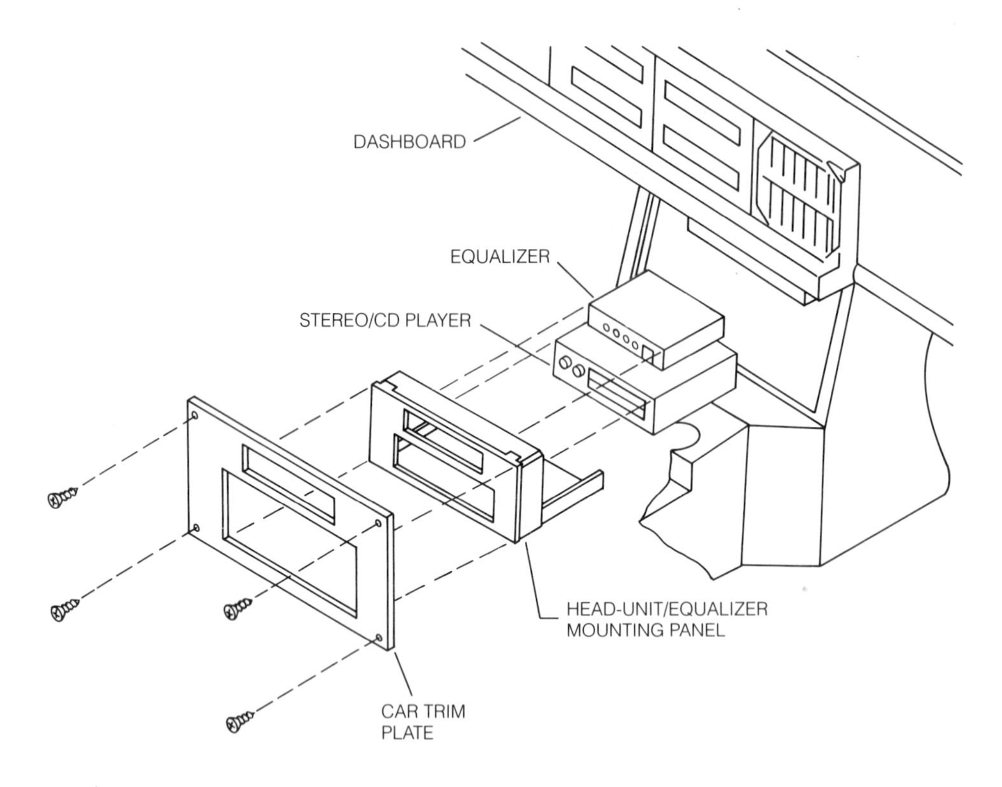

Figure 6-3. This installation kit permits in-dash mounting of both a DIN-E head-unit and a slim-line equalizer.

CAUTION

As discussed in Chapter 2, if you choose external amplifiers, do not overpower your speakers.

For our Pontiac Firebird example, we chose to leave the factory in-dash speakers, but drive them with a separate amplifier. We replaced the rear seat side-panel speakers and added subwoofers. The rear panel speakers are driven with a new power amplifier, as are the subwoofers.

If your plan contains a subwoofer, then the hatchback style automobile requires the use of a fabricated enclosure. You can buy prefabricated boxes at automobile sound stereo shops, but you can save quite a bit of money by constructing one yourself. In the Pontiac Firebird, we made use of the hatchback cargo well to house the subwoofers. This space provided an acoustic resonance box with approximately the correct volume and, instead of having to build a complete box, we only had to add a top. This took approximately 1/5 of the amount of wood it would have taken to build a complete box. We cut the top from one-quarter of a 4′ × 8′ sheet of 3/4″ dense particle board. It's importance that the particle board be the satin surface dense type. Grade A-B "void-free" plywood works just as well. Solid wood pieces of acceptable quality and size are not readily available and are expensive.

The details and dimensions of this top are shown in *Figure 6-4.* Be careful when measuring the curved sections and laying it out on the wood so that the top will fit in the cargo well neatly. Check your cargo area before you cut to make sure the dimensions are correct. After you make all initial cuts, lay the speaker board in place to check for fit. You can make slight modifications with a wood rasp to ensure a tight fit. Covering the wood with matching carpet gives a professional look to the installation. Often you can find carpet to match your automobile's interior in the automotive section of major department stores. You can go to your automobile dealer's parts department and usually find matching carpet, but you will probably pay more for it.

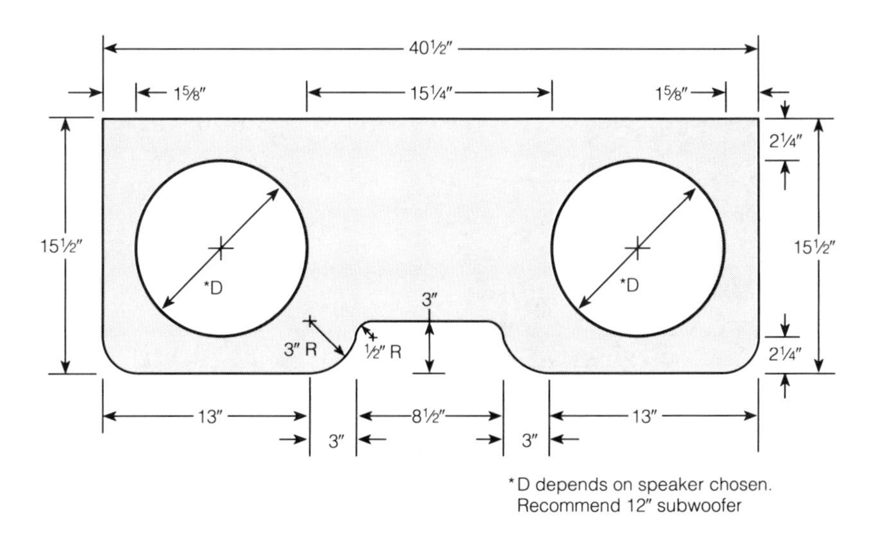

Figure 6-4. Dimensions for the top of the subwoofer enclosure formed by the Firebird's cargo well.

After carpet is applied, place the board over the well where it is to be mounted. We used self-drilling screws to secure the speaker board in place. *Figure 6-5* shows the speaker board anchored in place in the Firebird. Notice the wires that will be used to connect the subwoofers to the amplifier are sticking out of the holes for the subwoofers. *It is very important to have an air-tight fit.*

Next, connect the wires (be sure to observe polarity) to each of the subwoofers. Place the subwoofers in the precut holes and anchor them in place with 3/4" hardened steel screws, or special coarse thread particle board screws.

Our plan for the Firebird included power amplifiers, so we installed them next. Mounting the amplifiers was a relatively simple task in the Firebird. If possible, the amplifiers should be out of the passengers' immediate eyesight. To accomplish this goal, they can be mounted under a seat, behind a seat, in the trunk, or anywhere it can get adequate ventilation for cooling. In this project, we mounted an amplifier in each of the side panels; one for the subwoofers and one for the rear seat side panel speakers. The wiring in *Figure 6-5* indicates where one amplifier was mounted. We mounted the third amplifier for the dash speakers under the passenger's seat. If you

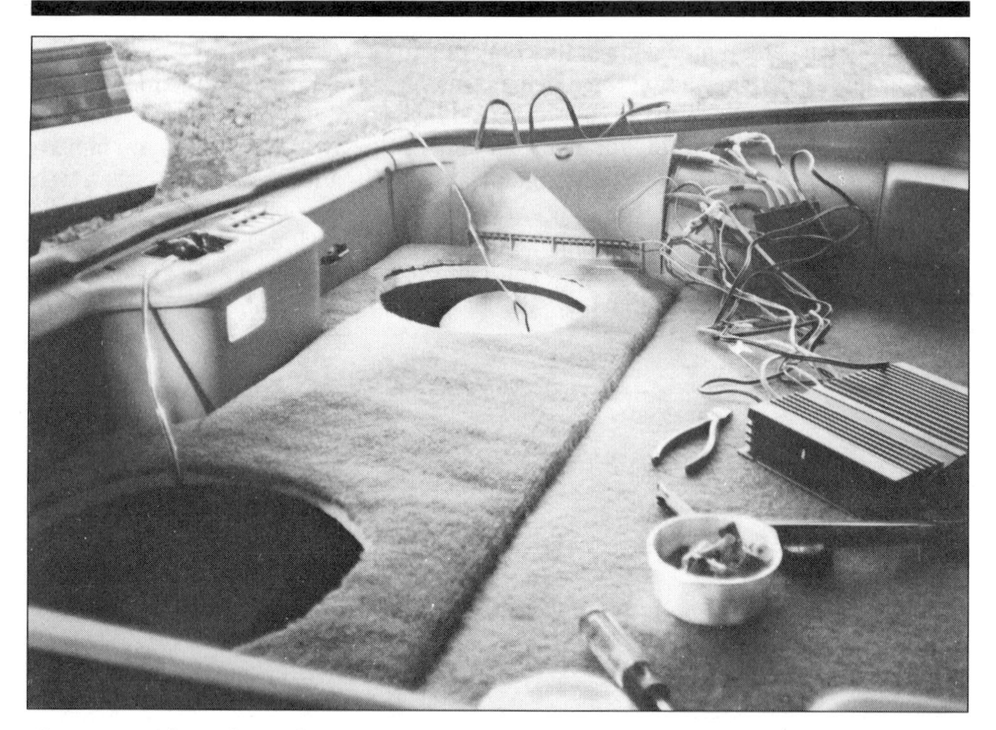

Figure 6-5. The subwoofer top is installed with wires in place to make the speaker connections. Wiring in the side panel indicates where the power amplifier will be mounted.

have a choice of mounting the amplifier under either the passenger's or the driver's seat, the passenger's seat is preferred because it is usually not moved as much as the driver's seat.

Most amplifiers come with a set of mounting brackets and mounting hardware. Usually, you can use four self-drilling screws to firmly secure the amplifier to the automobile's chassis. Refer to *Figure 5-4* for a mounting example.

CAUTION

If the amplifier is not securely mounted, it will probably move around. Stress on the wires and components due to this motion may cause wire breakage and/or permanent damage to the amplifier. A loose amplifier could become a dangerous missile when the vehicle comes to a sudden stop.

The total wiring diagram for this system is shown in *Figure 6-6*. The wires from the equalizer to all three of the amplifiers, the power lines for the amplifiers, and the rear seat side panel speaker wires are run under the floor carpeting; first to the amplifier under the passenger seat and then to the two amplifiers in the cargo-well side panels. Refer to *Figure 3-6* and *3-7* if any special cables or connectors are required.

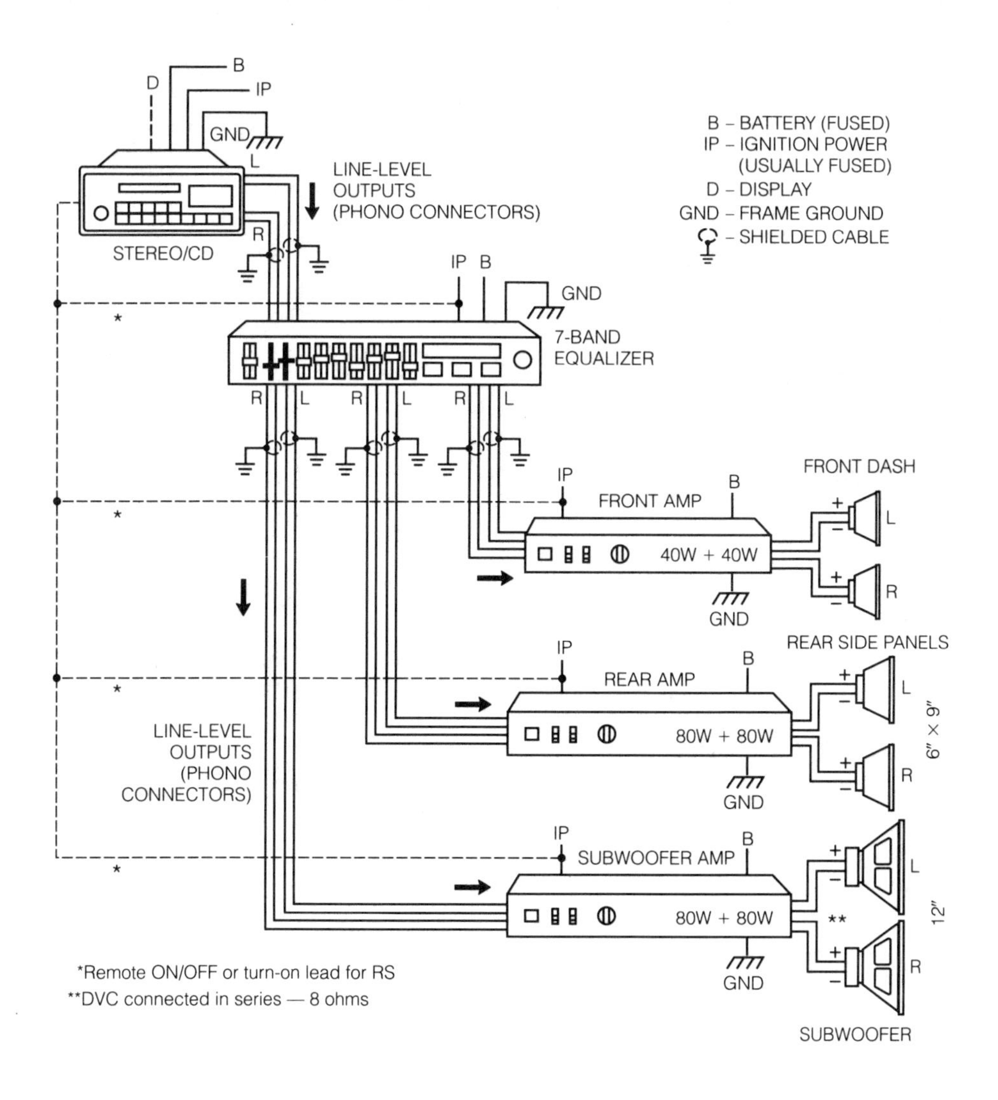

Figure 6-6. Wiring diagram for the sound system installed in a 1989 Pontiac Firebird.

Summary of System Performance

In addition to getting the nearly perfect sound features that we had set as our goal for this project, the system has a handsomely styled appearance that matches the sports car lines of the Firebird. *Figure 6-7* shows the subwoofers installed. They were covered with wire-mesh grilles to protect them from physical damage. Notice that the power amplifiers are completely out of sight in the side compartments.

Figure 6-7. Subwoofers are fastened in place. Grills will be added later. Power amplifiers are mounted out of sight in the side panels.

AN ECONOMY SYSTEM

For our second example in this chapter, we will describe an economy system designed and installed in a 1987 Toyota Corolla hatchback. We planned a system to meet the following specifications:

- An AM/FM stereo receiver with a digital electronic tuner, seek and scan, and automatic storage of station selections in its memory.
- A high-quality cassette tape player.
- A wide-range frequency response with a good fidelity sound.
- An ability to shape the sound frequency response to match the speaker system to the hatchback's interior and the listener's audio tastes.
- To accomplish these specifications for an investment of only about $300.

To meet our desired specification characteristics, we selected these components for this installation:

Item
DIN-C (double-shaft) AM/FM stereo with cassette player
7-band stereo equalizer/booster with speaker-level input
Pair of 5¼″ 3-way replacement speakers
Pair of 4″ full-range replacement speakers
DIN-C stereo/cassette installation kit (Toyota)
Wire, connectors, cables, hardware, etc.

CAUTION

Before attempting any installation, you should disconnect the negative terminal connection from the automobile battery to prevent damage to the vehicle wiring or the equipment you are installing.

Here is how we installed the system in the Toyota Corolla hatchback. First, the head-unit was installed in the existing opening left in the dash when we removed the factory AM/FM receiver. We compared the measurements from the Corolla's dash to the installation dimensions given for this stereo cassette unit. We determined that this unit would fit the mounting holes without modification of the dash by using an American International installation kit for a 1987 Toyota. The kit is sold in outlets that install automobile audio equipment.

Next, we mounted the equalizer under the dash as shown in *Figure 3-5*. The head-unit chosen only has speaker-level outputs. The equalizer chosen will accept speaker-level inputs so the units couple together directly. We ran the speaker wires for the rear speakers under the door threshold as shown in *Figure 3-10*.

Refer to Chapter 4 to see the detail of how we replaced the existing speakers with our selected upgrades. Both the 4" dash speaker replacements and the 5 1/4" side panel speaker replacements fit exactly into the holes left when we removed the factory speakers.

Figure 6-8 illustrates the wiring diagram for this system. Refer to *Figures 3-6* and *3-7* if any special cables or connectors are required. When it is completed, you will be ready to enjoy your new system.

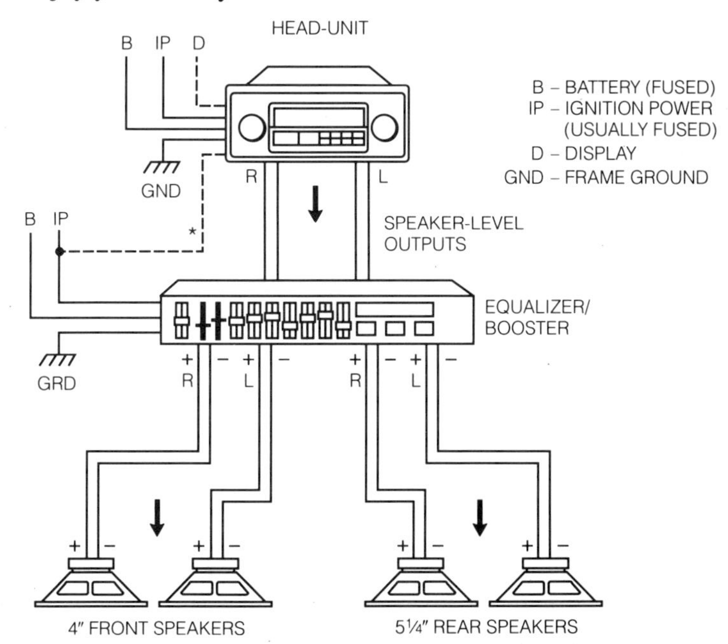

Figure 6-8. Wiring diagram for the sound system installed in a 1987 Toyota Corolla.

Summary of System Performance

In addition to adding the cassette player feature, the equalizer provides adjustment of the sound frequency response contour. This feature permits matching the interior of the smaller Toyota Corolla hatchback to give an "audio appearance" of a much larger space. In addition, the power output was increased by a factor of 4. For an economy system, the results were surprisingly good fidelity.

Pickup Truck Systems 7

The pickup truck is no longer used solely as a farm or construction vehicle. Its popularity has continually risen over the past decade and many people now drive pickups as their sole means of transportation. Pickup truck manufacturers have greatly improved the trucks' previously primitive interiors and exteriors so that they compare favorably to some of the nicer passenger cars on the road today. However, some people who use pickup trucks for everyday travel are not satisfied with factory-installed audio equipment and want to modify or replace the existing equipment.

Of all of the vehicle styles discussed in this book, the pickup is probably the most challenging to enhance because of the limited space. Constructing a speaker enclosure for woofers that are large enough to produce good bass and will fit in the available space is often difficult. An additional problem is that power amplifier mounting is usually limited to the space under the pickup seat. Keep these things in mind as you consider how you will upgrade your pickup's sound system.

Because there are hundreds of different combinations of makes and models, it is practically impossible to give step-by-step instructions for each. However, you should be able to apply the general information in this chapter to your specific installation.

SYSTEM PERFORMANCE GOALS

As with the coupe/sedan and the hatchback installation that we discussed in Chapters 5 and 6, your first goal in your stereo project should be to plan and establish what you expect from the completed system. Review Chapters 5 and 6 for guidelines for your system and Chapters 1 through 4 for mounting and wiring details that are required for the installations in this chapter.

We will describe two pickup installations in this chapter. The first is a "killer" system in a full-size 1984 Chevrolet Silverado.® The second is an economy system in a compact 1984 Ford Ranger.®

CAUTION

Before attempting any installation, you should disconnect the negative terminal connection from the automobile battery to prevent damage to the vehicle wiring or the equipment you are installing.

FULL-SIZE PICKUP INSTALLATION

The 1984 Chevy pickup is shown in *Figure 7-1*. Because the owner had taken good care of his truck and it looked practically new, he wanted a system that had superior sound specifications with minimal alterations to the truck's interior. Among his desired goals were enough power and low frequency response to pound out deep, clean, and undistorted bass notes. For the head-unit, he wanted an AM/FM stereo receiver with cassette player that had digital electronic tuning, a large number of FM presets, automatic station storage, memory-scan sampling, and LCD tuning/clock display. Also, because the pickup owner had a very large audio cassette music library, he wanted the cassette player to have Dolby B noise reduction, auto-search, metal/chrome selector, locking fast-forward, and key-off release.

In addition to the cassette tape format, he wanted to be able to plug in the remote-controlled portable CD player that he already owned. A universal CD swivel-mount bracket would be needed to secure the CD player and provide anti-skip damping action.

To meet the requirements of our design specifications, we selected the following components for this installation:

Item
DIN-E (flat-face) AM/FM stereo with cassette player
7-band passive stereo equalizer with subwoofer crossover
One 80-watt (40W per channel) stereo power amplifier
Two 160-watt (80W per channel) stereo power amplifiers
Two dynamic horn tweeters
Pair of 6″ × 9″ 3-way flush-mount speakers
Two 8" subwoofers with dual voice coils (for enclosure)
CD swivel-mount bracket (for portable CD player
Pair of speaker grilles
Two terminal plates for speakers
Material for enclosure
Wire, connectors, cable, hardware, etc.

CAUTION

Always wear safety glasses when drilling, sawing and filing to prevent eye injury from flying particles. Be sure to check behind the cutting area for wires or anything that could be damaged by drill bits or the saw blade. Remember, a sabre saw or jigsaw blade extends back several inches due to its reciprocating action.

Installing the Head-Unit

The DIN-E replacement head-unit is shown in *Figure 7-2*. Since the factory-installed AM/FM radio was a double-shaft style, we had to modify the dash so we could install the DIN-E head-unit. Included with a DIN-E head-unit is either a template or a specification sheet that gives precise measurements for the head-unit hole. Using the

Figure 7-1. This is the 1984 Chevrolet Silverado pickup truck in which the audio system was upgraded.

Figure 7-2. A DIN-E (flat-face) AM/FM stereo with auto-reverse cassette tape player. *(Courtesy of Panasonic)*

template, we carefully marked the area to be cut out and used a jigsaw with a metal cutting blade to cut the hole. The hatchback system installation of Chapter 6 required the same type cutout (refer to *Figure 6-1* for details). We followed the manufacturer's instructions and the details given in Chapter 3 for securing and wiring the head-unit.

Installing the Equalizer

The equalizer installation is the same as for Chapter 6. You may choose to install one with a bracket under the dash as shown in *Figure 3-5,* or an in-dash unit with an installation kit that accepts a DIN-E (one-half size) slim-line equalizer. In this installation, the equalizer is installed with a bracket under the dash. In Chapter 6, the DIN-E installation was used (see *Figure 6-3*).

Remember, an equalizer can have either line-level or speaker-level inputs. In this installation, we used the line-level inputs. The equalizer's line-level outputs were used to feed the power amplifiers.

Installing the Power Amplifiers

We chose a 40W/channel amplifier for the tweeters, an 80W/channel amplifier for the rear side quarter panel full-range speakers, and the same amplifier for the subwoofers. A power amplifier can be mounted under a seat, behind a seat, or anywhere it can get adequate ventilation for cooling (preferably out of the passengers immediate eyesight). Under the seat is usually the best place in a pickup truck. This is where we mounted the amplifiers in the Chevy pickup, as shown in *Figures 7-3a* and *7-3b.*

Many people think that removing the seat is difficult, but this is not necessarily true. Usually, pickup seats are held in place by four bolts located under the seat. After removing these bolts and pulling the seatbelt buckles down through their recessed openings, the seat can be lifted completely out of the pickup cab. Soliciting a friend's help to accomplish this task is wise. The seat is usually not very heavy, but it is rather bulky and may be hard to handle alone.

CAUTION

If the amplifier is not securely mounted, it will probably move around. Stress on the wires and components due to this motion may cause wire breakage and/or permanent damage to the amplifier. A loose amplifier could become a dangerous missile when the vehicle comes to a sudden stop.

We used four self-drilling screws to secure the amplifier, with its mounting brackets, to the vehicle's chassis. *Figure 7-3a* shows how the brackets are used to locate the holes for the self-tapping screws. We ran the wires from the equalizer to the amplifiers under the floor carpet by the door molding. Review Chapter 3 again, if necessary.

CHOOSING THE SPEAKER SYSTEM

Frequently, pickups have only two factory speaker openings in the dash. Some others come with speaker openings in the doors. A few newer trucks have speakers openings in the rear side quarter panels of the cab. As always, we recommend replacing any factory speakers with high-quality, full-range speakers. If four factory speaker openings are available, as was true in the 1984 Chevy pickup, be sure to replace them all. Failure to do so results in a one-sided, single-direction sound often referred to as "tubular." These speakers can be connected to the head-unit's speaker outputs or external amplifiers.

CAUTION

As discussed in Chapter 2, if you choose external amplifiers, do not overpower your speakers.

a. Installing Self-Drilling Screws

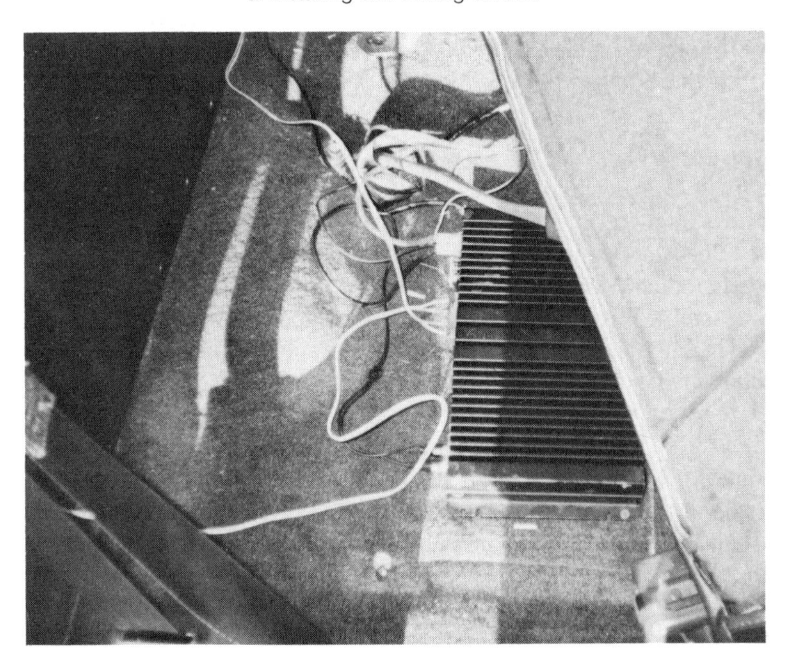

b. Amplifier in Place

Figure 7-3. The power amplifiers were mounted under the seat.

Using Subwoofers

If your plan contains subwoofers, as did our plan for the Chevy pickup example, then you will need to use fabricated speaker enclosures of some sort. If you build your own enclosure, you can save some money and have the satisfaction of doing it yourself.

Whether you build or buy your subwoofer enclosure, you will first need to measure the available space behind the seat. Here is how to do it: With the seat in its normal position, measure the space between the back wall and the seat back at the bottom and top of the seat back. Next, measure the height from the floor to the top of the seat back. Finally, measure the width of the cab's back wall. This will give you a good idea of the amount of available space and help you determine the maximum dimensions of your enclosure. The dimensions for the speaker enclosure for the 1984 Chevy pickup are given in *Figure 7-4a.* You may need to modify these to customize the enclosure to your particular truck. Because of the limited space, the enclosure resonance may have a peak which will emphasize the response to frequencies in the 50-100Hz range while reducing the response to frequencies below 50Hz. Therefore, keep the enclosure as large as possible. However, as pointed out in Chapter 5, the effective volume of an enclosure may be increased by 20% by filling it with Dacron pillow fill. See Chapter 5.

When you have the dimensions for your enclosure, purchase the wood necessary for the project. Cut (or have cut) the pieces to your dimensions using the layout guide shown in *Figure 7-4b.*

CAUTION

You can always cut a little more if the speaker hole is too small, but it is practically impossible to replace material if you cut the hole too big. Be careful!

Now you are ready to cut the speaker holes in the enclosure's speaker board. Use a template if one comes with the speaker. If not, *Figure 5-5* shows the appropriate way to measure the diameter of a speaker to obtain the speaker opening size. *Figure 4-12* shows how to cut the opening using a power sabre saw.

The next step is to fasten the box together. You can start with any two adjacent sides. Dry assemble all parts together by drilling necessary clearance and pilot holes as discussed in Chapter 5. Make sure all parts fit together properly. Apply a coating of yellow aliphatic resin glue (see Chapter 5) to one edge of each panel and secure the panels with hardened steel screws. Battens or blocks may be used as shown in *Figure 5-8* to reinforce the joints. After all six sides of the box are secured together, reach inside through the speaker holes and caulk all joints as shown in *Figure 5-9.*

Cut openings for two quick-connect terminal plates as shown in *Figure 5-10.* These terminals provide quick, easy and secure connections for the speaker leads from the amplifier. These terminals may be located in each side or in the front of the enclosure, depending on where it is more convenient. Once all cuts and holes are completed, use a rasp and sandpaper to smooth the rough edges.

Finishing the Enclosure

Although the enclosure will be out of sight, you still may want to apply carpet to the enclosure's exterior for a finished appearance. Carpet remnants can be purchased at most home carpet stores. If you want an exact match to the truck's factory-installed carpet, check at an automobile dealer's parts department. Cut pieces to fit each of the exterior walls of the box and use carpet mastic (for a neater job) or a staple gun to fasten them to the box. Use a single-edge razor blade or utility knife to cut openings for the speakers and the quick-connect pad.

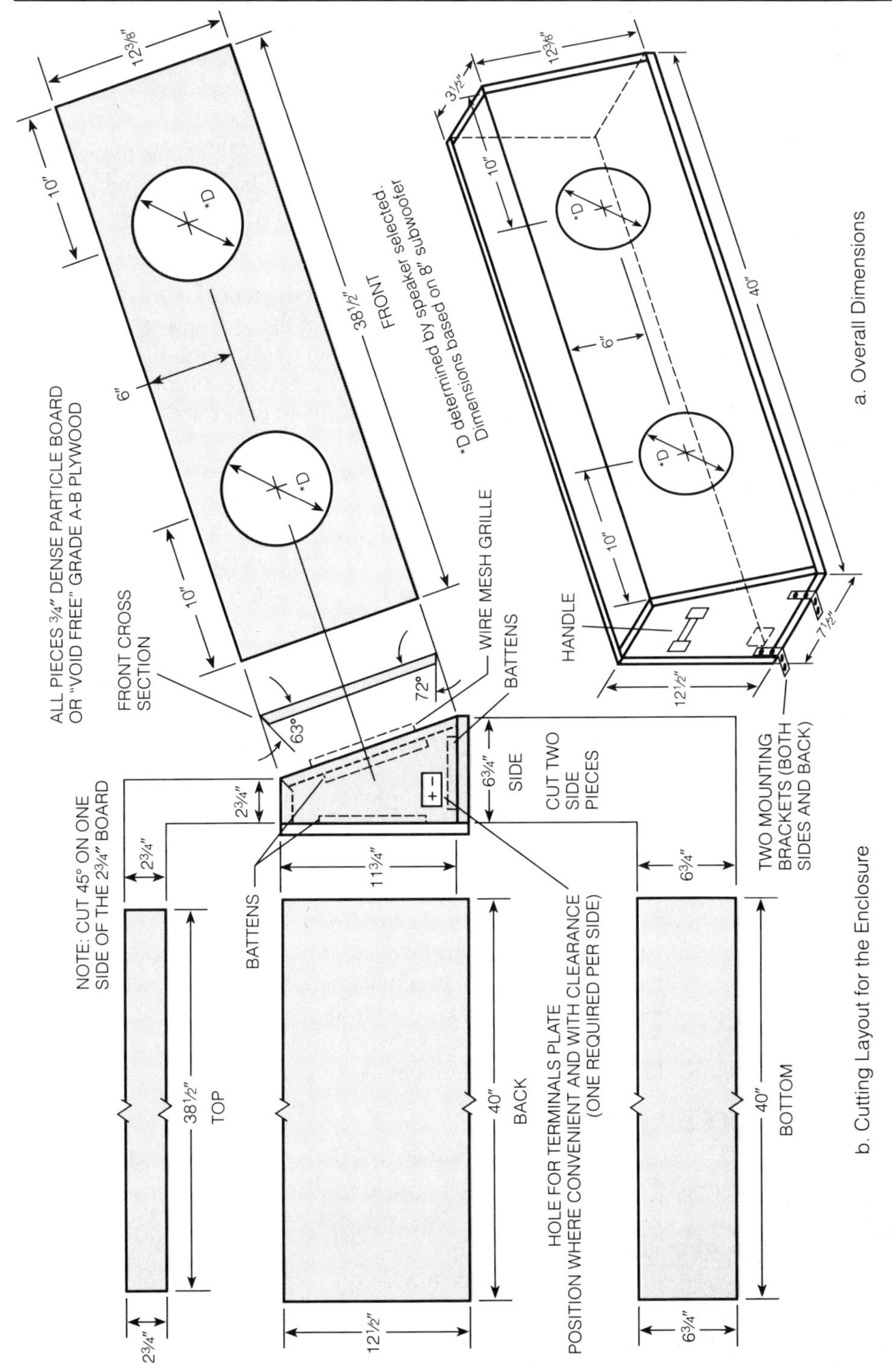

Figure 7-4. Cutting layout for the Chevy behind-the-seat enclosure.

Insert wires into the speaker openings, pass them through the inside of the enclosure, and out through the quick-connect terminal plate opening. Connect the appropriate ends to the speaker terminals (dual voice coils in series) and the inside terminals of the quick-connect pad. Be sure to keep polarity correct. Solder these connections. Stuff the wires back into the box and install the quick-connect pad and the speakers to the box with 1″ hardened steel screws. Finally, install speaker grilles over the speakers.

Mounting the Enclosure

The next step is to secure the enclosure in place. Fasten small “L” brackets to the sides at the bottom edge of the box. Place the box behind the seat and push it against the back wall of the cab. Anchor it in place with metal screws through the L brackets. Connect the speaker wires from the amplifier to the quick-connect terminal plate. Be sure to route all wires under the seat so that the forward and backward adjustments of the seat position will not damage any of the wires.

COMPLETING THE INSTALLATION

The schematic diagram for the system that we installed in the 1984 Chevrolet pickup is shown in *Figure 7-5.* Refer to *Figures 3-6* and *3-7* if any special interface cables or connectors are required. The bracket for the portable CD player was mounted to the floor in the middle of the truck in an upright position. Its output lead plugs into the equalizer input jack.

Route the power lines to all the amplifiers. Make sure there are good grounds to the pickup body (frame). Route the tweeter speaker leads from the dash openings, the input leads from the equalizer, and the power leads through door channel grooves and underneath carpeting to keep them hidden and out of the way. Check all wiring and system performance before putting the seat back in. This calls for reconnecting the vehicle battery, but only temporarily.

When everything checks out satisfactorily, disconnect the vehicle battery. After reinstalling the truck seat, reconnect the vehicle battery again. Your finished installation should look similar to the one shown in *Figure 7-6.*

The system performance of this pickup installation was simply great. It met the desired superior sound specifications with minimal alterations to the truck's interior. The system cranked out enough power and low-frequency response to pound out deep, clean and undistorted bass notes as the owner had requested. It provided the fixed input of the AM/FM and the portable input of the CD.

ECONOMY PICKUP INSTALLATION

For the economy system example, we chose a compact-sized 1984 Ford Ranger that had no sound system. We helped the owner come up with this list of goals for his new sound system.

- Good quality at minimum cost
- AM/FM stereo with digital electronic tuner
- Cassette player with locking fast-forward/rewind
- Full-range adjustment of frequency response
- Fader control for front/rear speakers

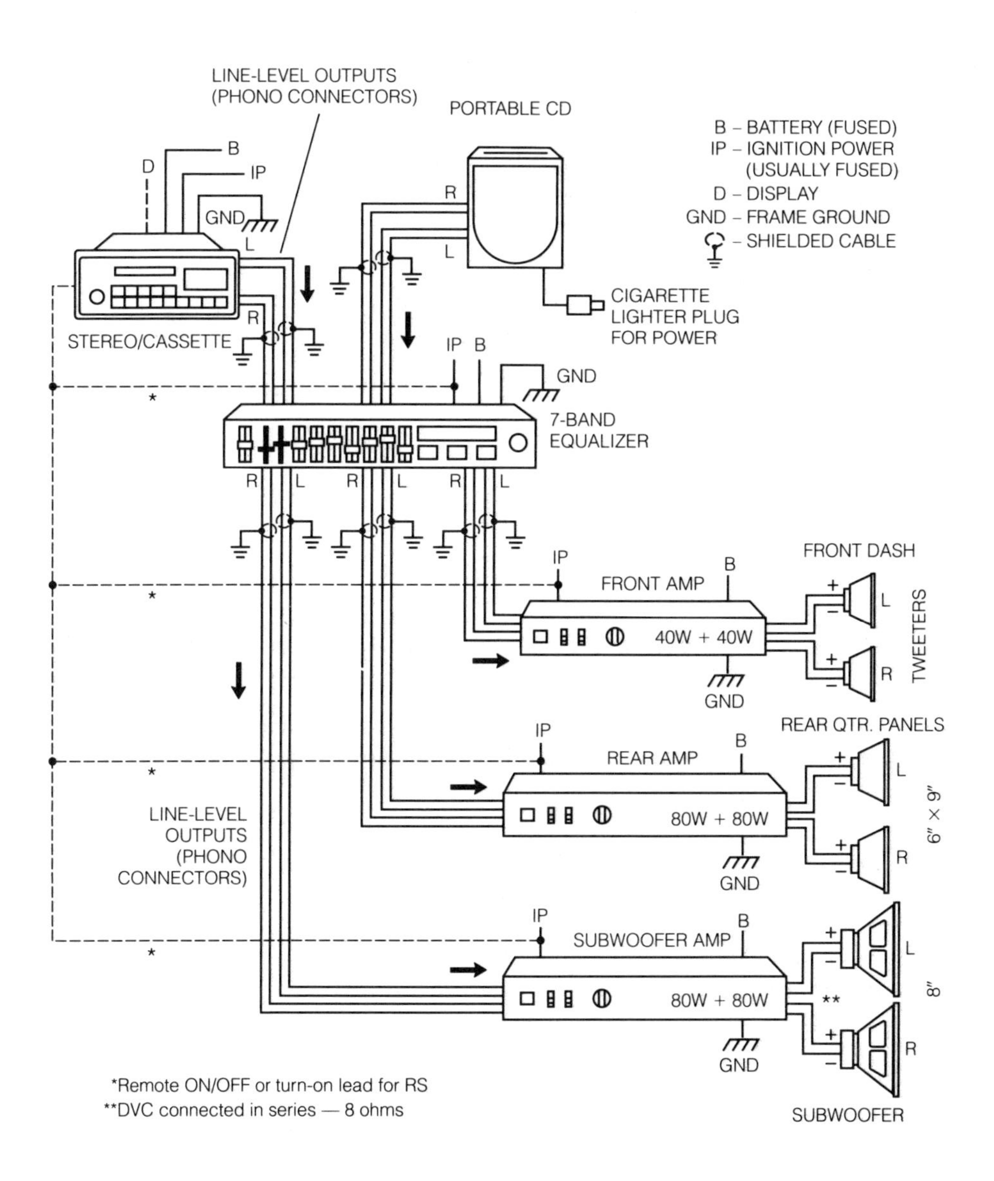

Figure 7-5. Wiring diagram for the sound system installed in the 1984 Chevrolet pickup truck.

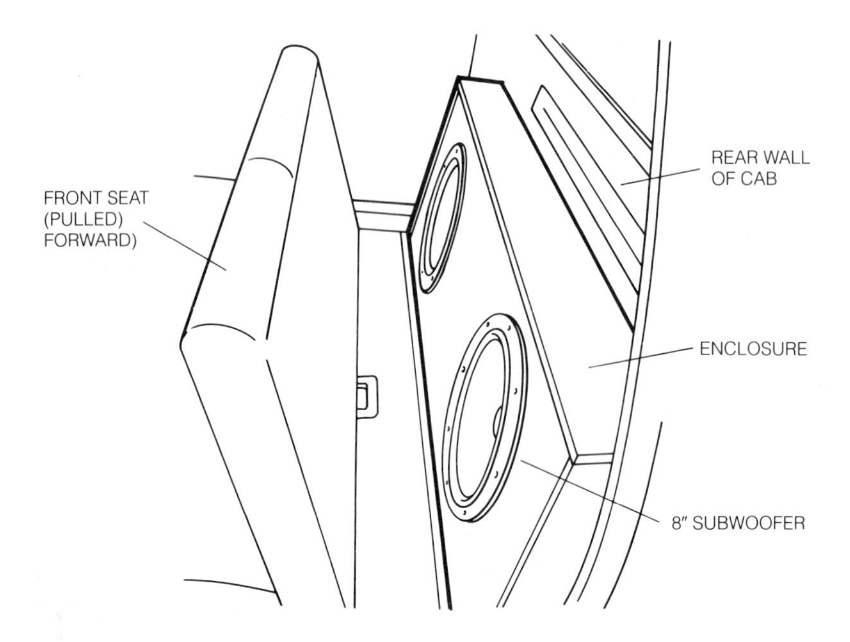

Figure 7-6. The fabricated enclosure with the 8″ subwoofers is installed behind the front seat of the 1984 Chevy pickup truck.

To meet these specifications, we selected the following components for this installation:

Item
DIN-E (flat-face) AM/FM stereo with cassette player
60-watt, 7-band equalizer/booster
Pair of 4″ full-range flush-mount dash speakers
Pair of prefabricated, carpeted, ported enclosures,
each having an 8″ 100-watt speaker and
top-firing piezo tweeter
In-dash DIN-E installation kit (Ford/Chrysler)
Car stereo antenna
Wire, connectors, cable, hardware, etc.

The DIN-E AM/FM stereo cassette head-unit, antenna and carpeted speakers are shown in *Figure 7-7.*

INSTALLING THE SYSTEM

Since this pickup never had a sound system, we had to remove the cover plate over the dash holes. An in-dash installation kit was used to securely mount the head-unit in the dash. There was over 8" of depth behind the dash for the head-unit, which was more than enough and made the head-unit installation easy.

a. Head Unit *(Courtesy of Kenwood)*

c. Prefabricated Carpeted Speaker Enclosures *(Courtesy of Kicker)*

b. Antenna *(Courtesy of Radio Shack)*

Figure 7-7. Economy system components for the 1984 Ford Ranger.

Since there was no existing antenna for the AM/FM receiver, we installed a one-piece 31″ stainless steel whip antenna. The first step was to knock out the plug which covered the antenna hole on the front fender. We inserted the antenna cable into the fender hole and fed the cable through the inner fender wall into the engine compartment (a large opening is available). We inserted the antenna base into the fender hole and secured the base to the fender.

CAUTION

Remember to check both sides of the firewall for obstructions before drilling. Pick an existing item, such as a bolt, that goes through the firewall as a reference to determine where to locate the hole.

Next, we drilled a hole through the firewall to feed the antenna cable into the passenger compartment. We selected a drill size to make a hole just large enough for the antenna cable connector to pass through. We fed the cable through the firewall hole, then slipped a rubber grommet onto the cable. We pushed the grommet into the firewall hole to protect the cable from the sharp edges of the hole. (An alternative method is to squeeze silicon sealant in the hole around the cable.) The final step was to plug the connector into the head-unit.

To drive the big woofers in the prefabricated truck boxes, we chose this equalizer/power booster for its economical features, selectable 4/8 ohm speaker output, and fader control to balance the 4-speaker system. We mounted the equalizer under the dash using its included bracket as shown in *Figure 3-5*.

The full-range 4″ flush-mount speakers were exact replacements, so they fit in the factory holes in the dash. We used the factory grille covers over these speakers. We placed the prefabricated speaker boxes behind the seat and anchored them with self-tapping screws and "L" brackets. We ran wires under the carpet from the equalizer to the rear speaker boxes. The schematic diagram for the system is shown in *Figure 7-8.* Refer to *Figures 3-6* and *3-7* if any special cables or connectors are required.

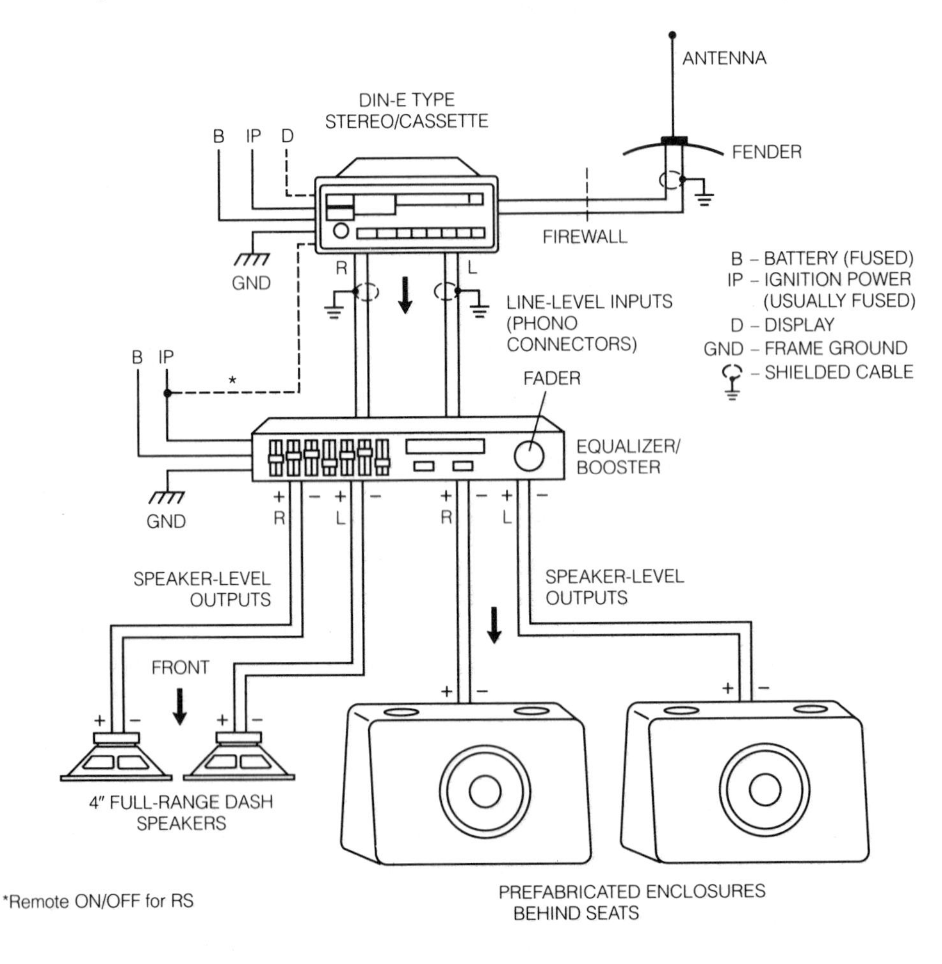

Figure 7-8. Wiring diagram for the sound system installed in a 1984 Ford Ranger.

SUMMARY

The system met our expectations and really had remarkable performance for the low cost (around $500). The pickup owner was happy with his new system.

As we pointed out in the beginning of this chapter, the pickup truck is probably the most challenging to enhance because of the limited space available. However, the two examples we presented show that a good-to-excellent system can be created, depending on how much time and money you want to spend.

Sport Utility Vehicle Systems 8

In this chapter, we will focus on the installation of sound systems in sport utility vehicles. Examples of sport utility vehicles are the Ford Bronco, Chevrolet S-10® Blazer, Toyota 4Runner,® and Jeep Cherokee.® This category includes any vehicle having two or four side-entry doors and one rear-entry door. This type of vehicle is similar in some ways to the hatchback, but for sound system modification, the sport utility is easier to work with because of its larger rear cargo area. Keep in mind that the larger volume of this area may require larger and more powerful speakers.

In many ways, the sport utility vehicle is also like the coupe or sedan, but a significant difference is the lack of a trunk in the sport utility. For subwoofer sound in the sport utility vehicle, a fabricated speaker enclosure is necessary.

We will present two systems as examples. The first example involved some innovative sound system engineering. In the second example, we used the existing factory system as a starting point and then enhanced it with "add-ons."

NOTE

Before beginning your installation, review Chapters 1 through 4 again. In Chapter 3, there are many mounting and wiring details given that are required in the installation.

A VERY HIGH-QUALITY SYSTEM

For our first example in this chapter, we will describe a top-notch system designed and installed in a 1988 Jeep Cherokee. We planned the system to meet the following specifications:

- An AM/FM stereo receiver with a digital electronic tuner, memory-scan sampling, seek, and automatic storage of station selections in its memory.
- A high-quality cassette deck with metal/chrome selection, auto stop, and locking fast-forward/rewind for ease of operation while driving.
- A full-range frequency response with a smooth, detailed sound contour from the crisp high tones all the way down to hard-hitting bass.
- A crossover to direct frequencies to the correct driver so the drivers can reproduce the audio spectrum without interference between drivers.
- Enough power from the system to overcome background road noise and to maintain the dynamic range of the signal from soft to loud so the music sounds live.
- The total system neatly installed using the space available in the sport utility vehicle, but with the components out of the view of the driver and passenger.

To meet our design specifications, we selected the following components for this installation:

Item
DIN-E (flat-face) AM/FM stereo with cassette player
7-band passive stereo equalizer with subwoofer crossover
One 160-watt (80W per channel) stereo power amplifier
Two 80-watt (40W per channel) stereo power amplifiers
Two 8″ subwoofers with dual voice coils (for enclosure)
Two pairs of 5¼″ replacement speakers
Pair of speaker grilles
Speaker terminal plate
Material for enclosure
Wires, connectors, cables, hardware, etc.

CAUTION

Before attempting any installation, you should disconnect the negative terminal connection from the automobile battery to prevent damage to the vehicle wiring or the equipment you are installing.

Installing the System

You may want to purchase all the components for your system to give yourself time to examine them before you begin the installation. Carefully read all of the manufacturer's instructions included with each component before beginning the actual installation. If an installation kit is required, then thoroughly read the instructions for the kit and how it is installed.

Installing the Head-Unit

The head-unit is shown in *Figure 8-1*. Its line-level outputs interface to the passive equalizer, which in turn, drives the power amplifiers. This head-unit has above-average FM tuner features with its 30 station presets. Its superior auto-reverse cassette player functions include locking fast-forward/rewind and a "key-off release." The key-off release is important because it protects the cassette tapes from being damaged if the ignition switch is turned off while the unit is in the "play" mode. With this feature, the tape drive motor reverses momentarily to release the tension on the tape and disengages the head and capstan.

Carefully follow the step-by-step instructions included with your new head-unit to install it securely. If you are using an installation kit, again follow the instructions carefully to mount it into the dash. Then mount the head-unit in the hole designed for it. Two kit installations are shown in *Figure 5-3* and *Figure 6-3.* If a template comes with the head-unit, hold the template in position and use it to mark the outer perimeter of the head-unit. Remember, if any part of an existing hole in the dash or the installation kit extends beyond the outer edges of the template, then the chosen head-unit is too small. You will have to re-examine your plan and either get a new head-unit or a new installation kit. If a new hole is required, cut the hole using a jigsaw or a sabre saw. The techniques and details are shown in *Figure 6-1.*

Figure 8-1. This is a DIN-E (flat-face) AM/FM stereo with cassette tape player.
(Courtesy of Coustic)

This head-unit, like many others, uses an anchor jacket. Slide the jacket into the dash hole and bend the anchor tabs so that they grasp the interior edges of the hole to secure the anchor jacket in place. Slide the wiring harness and the head-unit into the jacket and make sure the self-locking mechanism closes securely. Follow the instructions carefully, so that when you snap the outer trim plate in place, it will be a neat installation.

An anchor strap is included with most head-units. We haven't discussed this before, but a properly attached anchor strap helps stabilize the head-unit installation and also can act as a theft deterrent. Attach one end of the strap to the bolt on the rear of the head-unit and the other end to the automobile chassis.

The head-unit requires battery power for its preset memory, ignition-switched power, display and ground connections for the main circuitry, and input and output signal connections. The system wiring diagram is shown in *Figure 8-2.* Remember, if you are mixing units from different manufacturers, special interface connections or cables may be required. Refer to *Figures 3-6* and *3-7* if special cables or connectors are required.

Installing the Equalizer

The equalizer we chose is the same one that we have used in several of the other installations. Let's review its features:

- It has wide-range passive tonal adjustment for all three of its outputs—the subwoofer amplifier, the front power amplifier, and the rear power amplifier.
- Its built-in selectable crossover frequency for the subwoofer output allows the crossover frequency to be either 60Hz or 100Hz.

These features permit continuous adjustment of the sound over a large dynamic range and wide frequency spectrum.

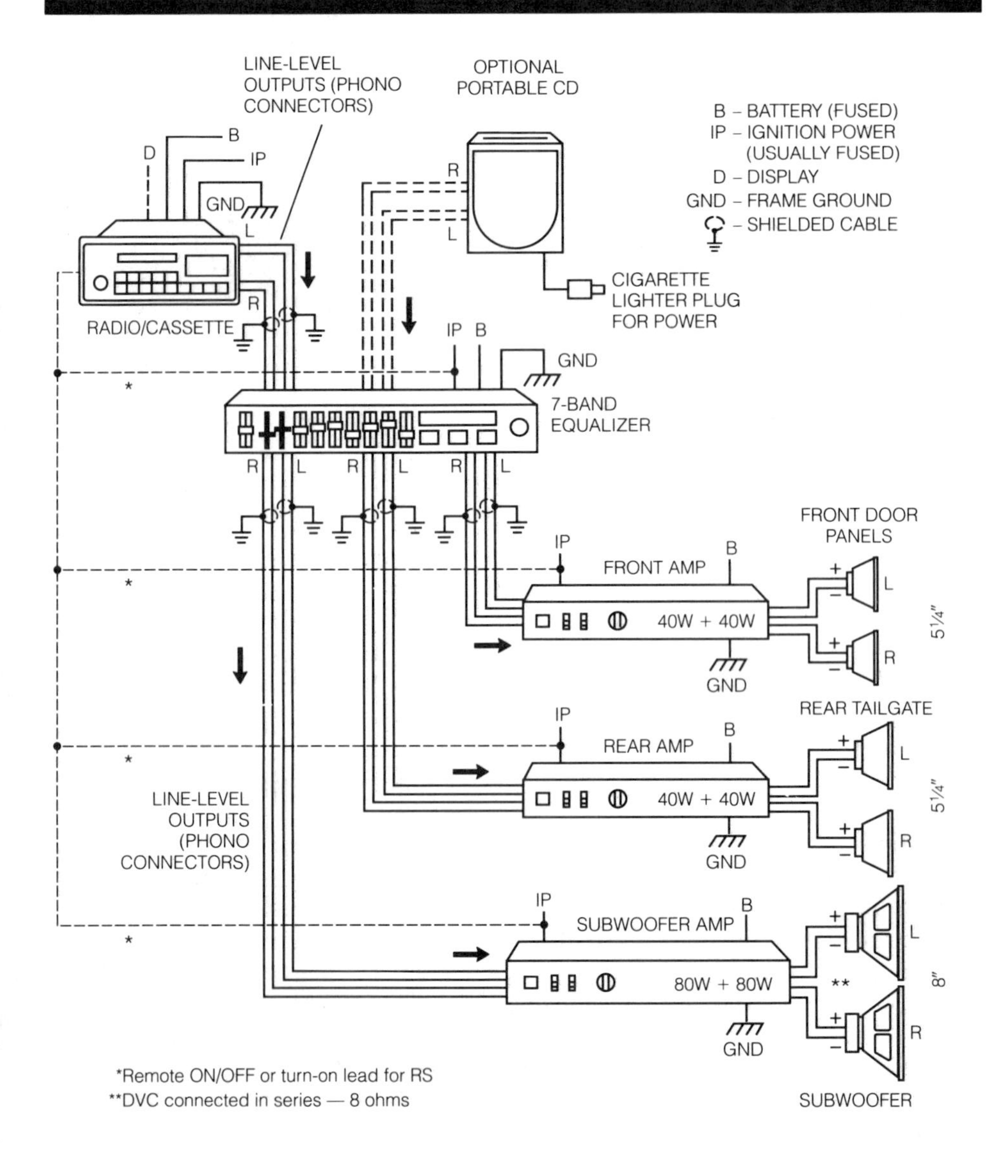

Figure 8-2. Wiring diagram for the sound system installed in a 1988 Jeep Cherokee.

As we mentioned before, some installation kits for the head-unit permit a slim-line equalizer to be mounted in the same adapter. We showed such an installation in *Figure 5-3* for a double-shaft installation kit and in *Figure 6-3* for a flat-face kit. If you want to use an under-dash bracket to install the equalizer, see *Figure 3-5* for details. If your system wiring is different from *Figure 8-2*, refer to Chapter 3 for wiring hookups that coordinate with your particular system. If you are mixing units from different manufacturers, refer to *Figures 3-6* and *3-7* for cables and connectors provided with certain units. Special connectors and cables are available from most electronics stores.

Installing the Amplifiers

For installation in this Jeep Cherokee, we chose to include two 80-watt stereo amplifiers and one 160-watt stereo amplifier. As shown in *Figure 8-2*, the 80-watt amplifiers drive speakers in the front door panels and speakers in the rear tailgate. The 160-watt amplifier drives the subwoofer speakers in a separate enclosure. Refer to Chapter 5 for the amplifier's specifications.

In this installation, as with the amplifiers in the hatchback system of Chapter 7, we mounted the amplifiers under the seats. *Figures 7-3a* and *7-3b* show the details. In a sport utility vehicle, there may be other locations where they can be mounted so that they will not be easily seen. Most places will be satisfactory as long as the amplifiers have adequate ventilation and can be mounted securely.

CAUTION

If the amplifier is not securely mounted, it will probably move around. Stress on the wires and components due to motion may cause wire breakage and/or permanent damage to the amplifier. Also a loose amplifier could become a dangerous missile when the vehicle comes to a sudden stop.

Run the power leads, the "turn-on" lead, the input leads, and the speaker leads under the carpet and in the channel paths under the threshold plate at the bottom of the door openings (see *Figure 3-10* for details). Use the phono connectors as indicated, and make sure each amplifier has a good ground to the vehicle chassis.

CAUTION

Do not connect the "turn-on" lead of the amplifier to a constant power source. If you do, the amplifier will be on continuously and run down your battery. Connect it to the remote on/off control line shown in Figures 3-6 and 3-7, or to an ignition-switched (IP) power source.

If your head-unit is equipped with a power antenna control line, the amplifier "turn-on" lead can be connected to the same lead that controls the antenna motor. If not, then it can be connected to a convenient ignition-switched power source. A 20-gauge wire is adequate for the "turn-on" lead because it will carry only a low current that controls a relay inside the amplifier. Power applied to the "turn-on" lead causes the relay contacts to connect the constant (battery) power source to the amplifier circuits. Remember, the battery power lead to the amplifier must be at least a 14-gauge wire to carry the high current drawn by the amplifier.

Speaker Installation

Replace the factory speakers with improved after-market upgrades. (If you carefully checked the replacement speakers before purchasing, then they should mount in the factory-provided holes.) Begin with the speakers located in the front lower door panels. Follow *Figure 4-5* for reference. First, remove the factory grille covers, being careful not to damage them because they will be used to cover the replacements. Usually four screws hold the factory speaker in place. Remove these screws and the speaker can be easily removed. Disconnect the speaker leads. Observe polarity and reconnect the leads to the replacement speaker. Position the new speaker in place, replace the screws, and tighten them for proper anchoring. Finally, reinstall the factory grille covers for a handsomely finished installation. Simply repeat this procedure for the speakers located in the rear tailgate.

Building a Speaker Enclosure

Figure 8-3a shows an enclosure with two front-mounted 8" woofers that will fit in almost any size sport utility vehicle. Measure the dimensions of the rear cargo area of your vehicle. See Chapter 5 if you have to modify the dimensions to fit your space. For subwoofers, try not to reduce the box *volume* to less than that in *Figure 8-3.*

When you have the dimensions for your enclosure, purchase the wood necessary for the project. Cut (or have cut) the pieces to your dimensions using the layout guide shown in *Figure 8-3b.*

CAUTION

You can always cut a little more if the speaker hole is too small, but it is practically impossible to replace material if you cut the hole too big. Be careful!

Measure the cutting line for the subwoofers chosen using a template that usually comes with the speakers, or refer to *Figure 5-5.* It shows the appropriate way to measure the diameter of a speaker to obtain the speaker opening size. Drill a small hole inside the cutout mark to start cutting the speaker hole. Cut the holes with a power saber saw equipped with a wood-cutting blade as shown in *Figure 4-12.* Set the speaker in the opening for a trial fit. Modify each speaker opening with a file until its speaker properly fits in the hole. Smooth the rough edges with sandpaper.

Assembling the Box

Start the box assembly with any two adjacent sides. Follow the procedure of dry assembling the box before gluing outlined in Chapter 5. Battens or blocks may be used as shown in *Figure 5-8* to reinforce the joints. After all six sides of the box are secured together, reach inside through the speaker holes and caulk all joints as shown in *Figure 5-9.*

As with the enclosures in previous chapters, the connections to the speakers inside the enclosure are made through two quick-connect speaker terminals as shown in *Figure 5-10.* These terminals provide quick, easy and secure connections for the speaker leads from the amplifier. An opening must be cut in the side of the enclosure for these terminals. The cuts are made with a jigsaw or sabre saw the same way as the speaker openings. Use a rasp and sandpaper to smooth the rough edges of the holes.

Since this enclosure is likely to be seen by passengers, it should be finished. As we discussed previously, covering the box with carpet is a convenient way to provide a neat appearance. See Chapters 5, 6 and 7 for details.

Completing the Wiring

Insert wires into the speaker openings, through the enclosure, and out the quick-connect pad openings. Connect the subwoofer dual voice coils in series (8 ohms — see *Figure 5-6b*). Solder each wire to a speaker terminal and the other end to an inside terminal of the quick-connect pad. Be sure to keep polarity correct. Stuff the wires back into the box and install the quick-connect terminals and the speakers to the box with 1" hardened steel screws. Finally, install speaker grilles over the speakers.

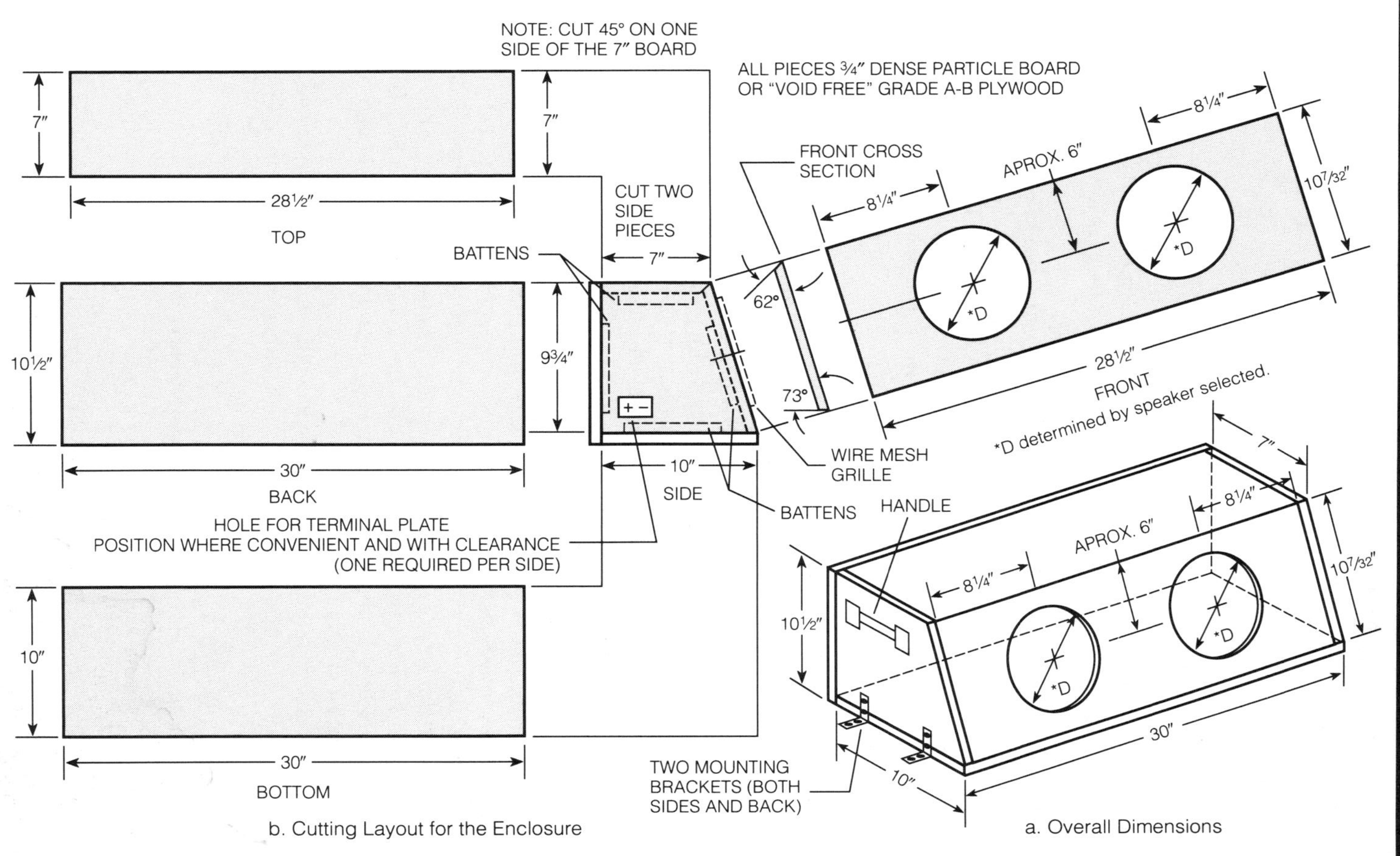

Figure 8-3. Speaker enclosure for a 1988 Jeep Cherokee.

Mounting the Enclosure

Fasten the enclosure in place with small "L" brackets at the bottom edge of the box as shown in *Figure 8-3a.* Place the box in the rear cargo area and anchor it in place with metal screws through the L brackets. Connect the speaker wires from the amplifier to the quick connect pad and the job is complete.

Wiring the System

In the wiring diagram of *Figure 8-2*, notice that the equalizer has an input to accept the output of a portable CD player. Most of the portable CD players sold have a cable included. However, if your portable CD player does not have a cable, you can obtain one from an electronics store.

Since the 80-watt amplifier output leads are connected to the door and tailgate speakers via the existing factory-installed wires that terminate in the head-unit area, wires must be run from the outputs of both 80-watt amplifiers to the head-unit area. Run these wires, and the wires from the equalizer outputs to the amplifier inputs, under the floor carpet at the side molding. Run the wires from the 160-watt subwoofer amplifier outputs to the speaker enclosure under the floor carpet along the edge of the side wall. Connect all wires to the appropriate connectors.

Use this installation suggestion when hiding the wires in a sport utility vehicle: Take a relatively small screwdriver and wrap a cloth around the end of it. Lay the wire in the corner of the carpet where the side wall and floor of the vehicle come together. Use the screwdriver to push the wire gently into the crevice. Be very careful not to cut or puncture the wire. Continue this until all the wire is neatly hidden under the carpet.

This completes the installation of the Jeep Cherokee system. The system has full-range frequency response with a smooth detailed sound contour from the crisp high tones all the way down to hard-hitting bass.

FACTORY "ADD-ON" SYSTEM

The next installation uses an existing factory system which is enhanced with "add-on" system components. Some of the newer (1985 to present) automobiles come equipped from the manufacturer with a relatively adequate four-speaker stereo system. Often these systems can be enhanced to a "very good" listening standard by the addition of low-end response. For this example, we enhanced a 1989 Chevrolet S-10 Blazer. This sport utility's factory system proved to have plenty of high- and mid-range sound, but was somewhat lacking in its bass response.

The "add-on" system for the Blazer included the following components:

Item

For "Add-on"

Two homemade speaker-level to line-level power converters
Two 12″ subwoofers with dual voice coils (for enclosure)
One 160-watt (80W per channel) power amplifier
Terminal plate for speakers
Material for enclosure
Wire, connectors, cable, hardware, etc.

<u>*Possible Expansion*</u>
Two flared-horn piezo tweeters
Two 2-way crossover networks

The subwoofer amplifier had line-level inputs, but the factory system had speaker-level outputs. Therefore, speaker-level to line-level converters had to be used between the factory system outputs and the subwoofer amplifier inputs. *Figure 8-4* shows such a level converter can be simply constructed. Two are required—one for each channel. Each converter requires two resistors connected as a voltage divider and a cable with a phono connector on one end and bare wires on the other end. The two cables can be easily made by purchasing a cable with phono connectors on both ends. Cut the cable in half and use one-half of the cable to make each level converter. Carefully strip the shield back about 1 1/2" to 2." Solder the shield to one end of resistor R2, which will connect to the minus (-) terminal of the speaker-level output of the head-unit. Solder the center conductor to the junction of resistors R1 and R2 (point "A"). Run wires from the minus (-) and plus (+) speaker-level output terminals on the radio (or from the existing speaker terminals) to the location of the subwoofer amplifier. At the subwoofer amplifier location, connect these wires (the speaker-level output leads) to the input of the level converter (point "B") by using crimp connectors or wire nuts. Connect the phono connectors to the line-level inputs of the subwoofer amplifier. In this installation, the amplifier is mounted under a seat just like the Jeep Cherokee installation. The amplifier feeds the speaker in the enclosure of *Figure 8-5*.

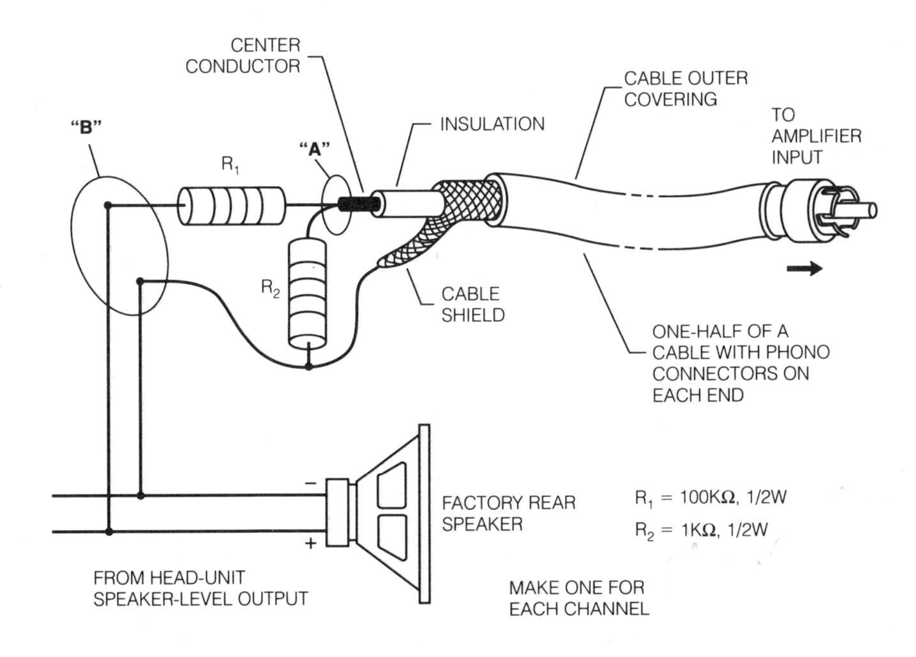

Figure 8-4. Construction of a speaker-level to line-level converter.

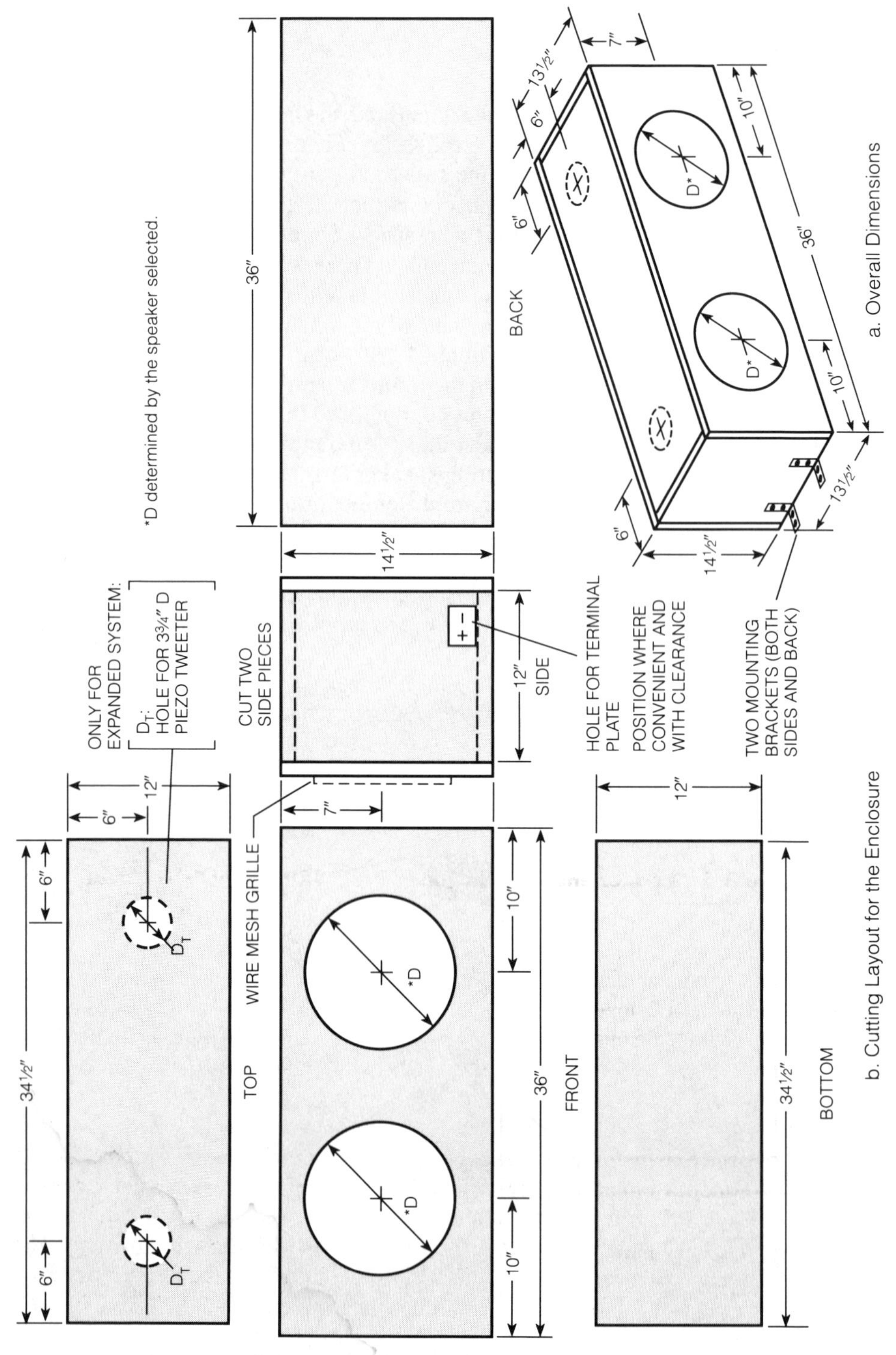

Figure 8-5. Speaker enclosure for a 1989 Chevy S-10 Blazer.

Speaker Enclosure

The overall dimensions of the speaker enclosure for this add-on system are given in *Figure 8-5a* and the cutting layout is shown in *Figure 8-5b.* This enclosure is designed to greatly improve the low and the very low bass as well as to more evenly distribute the high frequencies. It is constructed and mounted the same as described above for the Jeep Cherokee. The finished enclosure installation, with the high-frequency enhancement discussed below, is shown in *Figure 8-6* looking into the Blazer from the rear door.

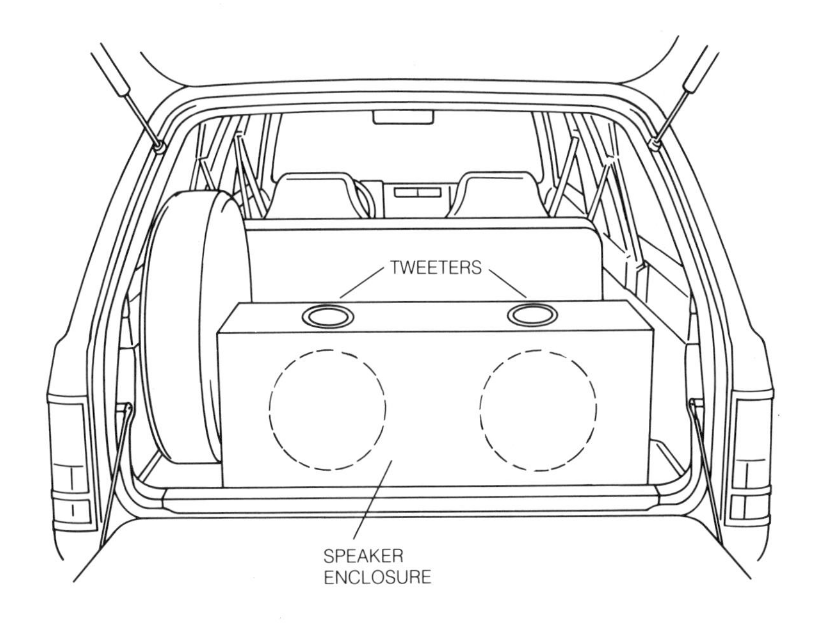

Figure 8-6. Speaker enclosure installed in a 1989 Chevy S-10 Blazer.

The system wiring diagram is shown in *Figure 8-7.* The wires were run and hidden as explained above for the Jeep Cherokee. The system provided greatly improved bass response over the original factory system. If the factory installed head-unit has a fader control, the volume of the add-on system will vary with fader position.

High-Frequency Enhancement

Another possible expansion is indicated on the enclosure drawing of *Figure 8-5.* If more high-frequency enhancement is desired, two tweeters can be added to the enclosure. When this is done, two crossover networks must be added to the enclosure speaker wiring. The diagrams on the crossover networks show how to wire the crossover networks to the speakers: The addition is indicated for the enclosure on the wiring diagram of *Figure 8-7.* The crossover frequency is 3000Hz. The dual voice coils of each subwoofer are connected in series (8 ohms).

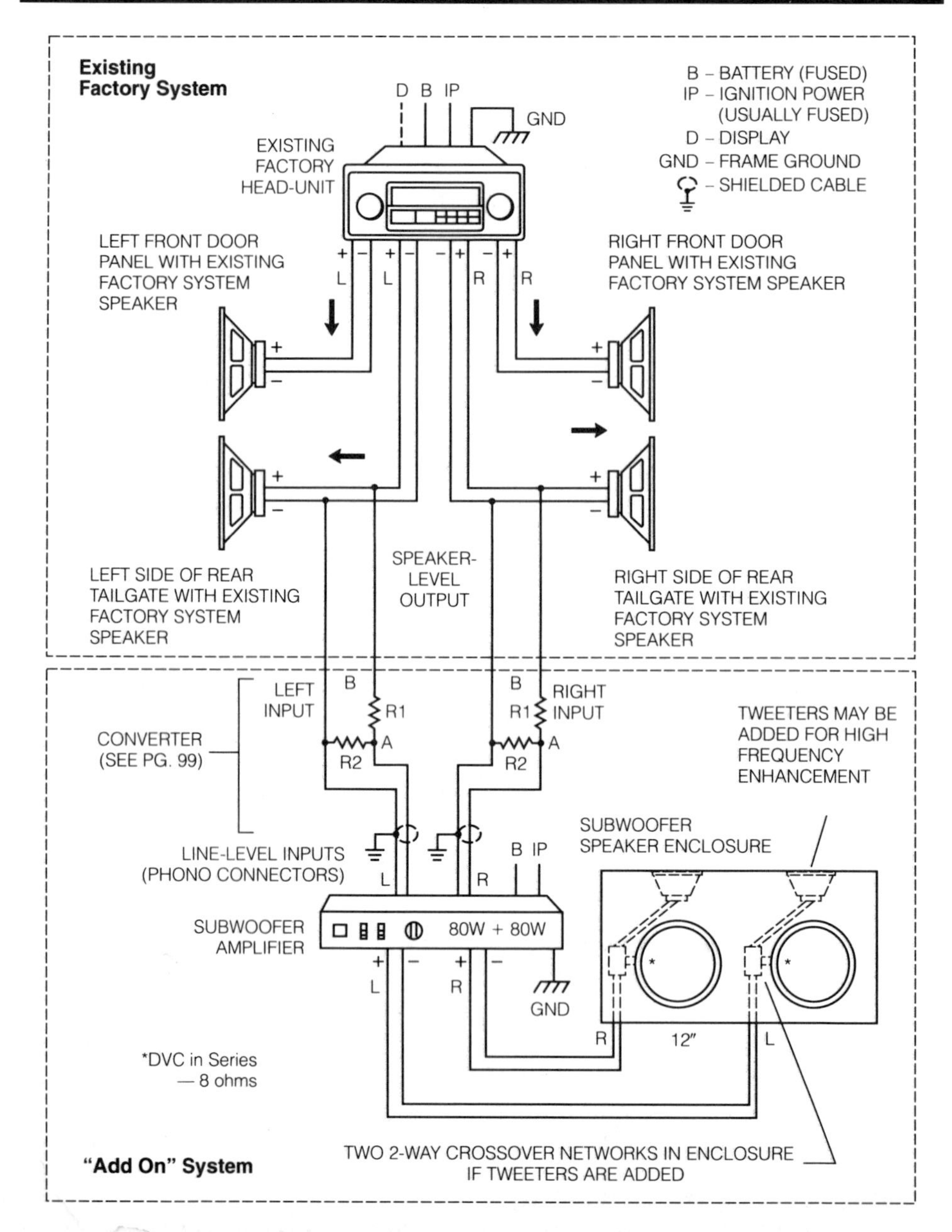

Figure 8-7. Wiring diagram for the sound system installed in a 1989 Chevy S-10 Blazer.

SUMMARY

Stereo modification is easier in a sport utility vehicle than most automobile styles because of its large rear cargo area. This large area does not limit the size of the enclosure and a true bass reflex enclosure resonant at a very low frequency can be designed. However, bigger and more powerful speakers may be required to fill the larger volume listening area.

VAN SYSTEMS 9

This chapter focuses on full-size vans and minivans. These vehicles usually have three side doors and two rear doors. In some ways, a van is similar to a hatchback, but stereo modification in a van is easier because of the large rear cargo area. However, because the interior of a van is usually open, it is sometimes difficult to hide the equipment and wires.

In this chapter, we will build and install two separate systems with independent sources and speakers. The front source is a stereo AM/FM receiver and cassette player head-unit and the rear source is a CD player. Since passenger vans are often used for extensive travel, two independent systems are an advantage because passengers often have different listening tastes. The system, while capable of working as two independent systems, is also capable of being phased together for a full surround-sound effect. This chapter includes the steps for the construction of a simple control box which permits the head-unit or the CD player, respectively, to be played throughout the system, or one can be played through the front speakers and the other through the rear speakers.

The 1988 Dodge RAM® van that we chose for this installation was equipped with a full after-market customized interior. It had full floor and wall carpeting, four captain's chairs, and a couch in the rear. The carpet and couch were helpful in hiding the equipment from the passengers.

PLANNING THE SYSTEM

To satisfy the top quality performance specifications desired, we chose the following components:

Item
DIN-C AM/FM stereo with cassette and 5-band equalizer
Under-dash CD player (for wall mounting)
Three 80-watt (40W per channel) stereo power amplifiers
Pair of 4-way surface-mount speakers (front)
Pair of 4-way surface-mount speakers (rear)
One 12″ woofer with dual voice coils (for enclosure)
Wire mesh speaker grille
In-dash installation kit (Ford/Chrysler)

Phono cables (lengths and quantities depend on the specific installation)
- 3′
- 6′
- 12′
- 24′

5-pin connector (male)
Two 4-position barrier strips
Three 5-pin connectors (female)
Two terminal plates for speakers
Homemade A/B control box
Material for enclosure
Wire, connectors, cables, hardware, etc.
* (Separate parts list is provided later)

Component Locations

Figure 9-1 shows the component mounting locations and the wiring of the power and ground connections. In the front of the van, we will use the in-dash mounting kit to install the head-unit. We chose this head-unit for its auto-reverse high-quality cassette player, its capability of storing 30 radio station presets, the included five-band equalizer for response adjustment, and its line-level outputs that will help link it with the rest of the system. The homemade A/B control box will be installed in the front close to the driver.

On the front doors, a pair of four-way surface mount speakers will be installed. We chose these speakers so they would fit without cutting new openings in the doors.

In the rear of the van, another pair of these same four-way surface mount speakers will be installed just in front of the rear fenders. The CD player will be mounted directly to the wall in front of the couch so rear passengers can operate it. Three identical power amplifiers will be installed under the couch; each is capable of supplying 40 watts per channel. A speaker enclosure which has one 12" woofer with dual voice coils also will be installed under the couch. This single woofer box will help achieve the complete surround sound that is specified.

NOTE

Before beginning your installation, review Chapters 1 through 4 again. In Chapter 3, there are many mounting and wiring details given that are required in the installation.

CAUTION

Before attempting any installation, you should disconnect the negative terminal connection from the automobile battery to prevent damage to the vehicle wiring or the equipment you are installing.

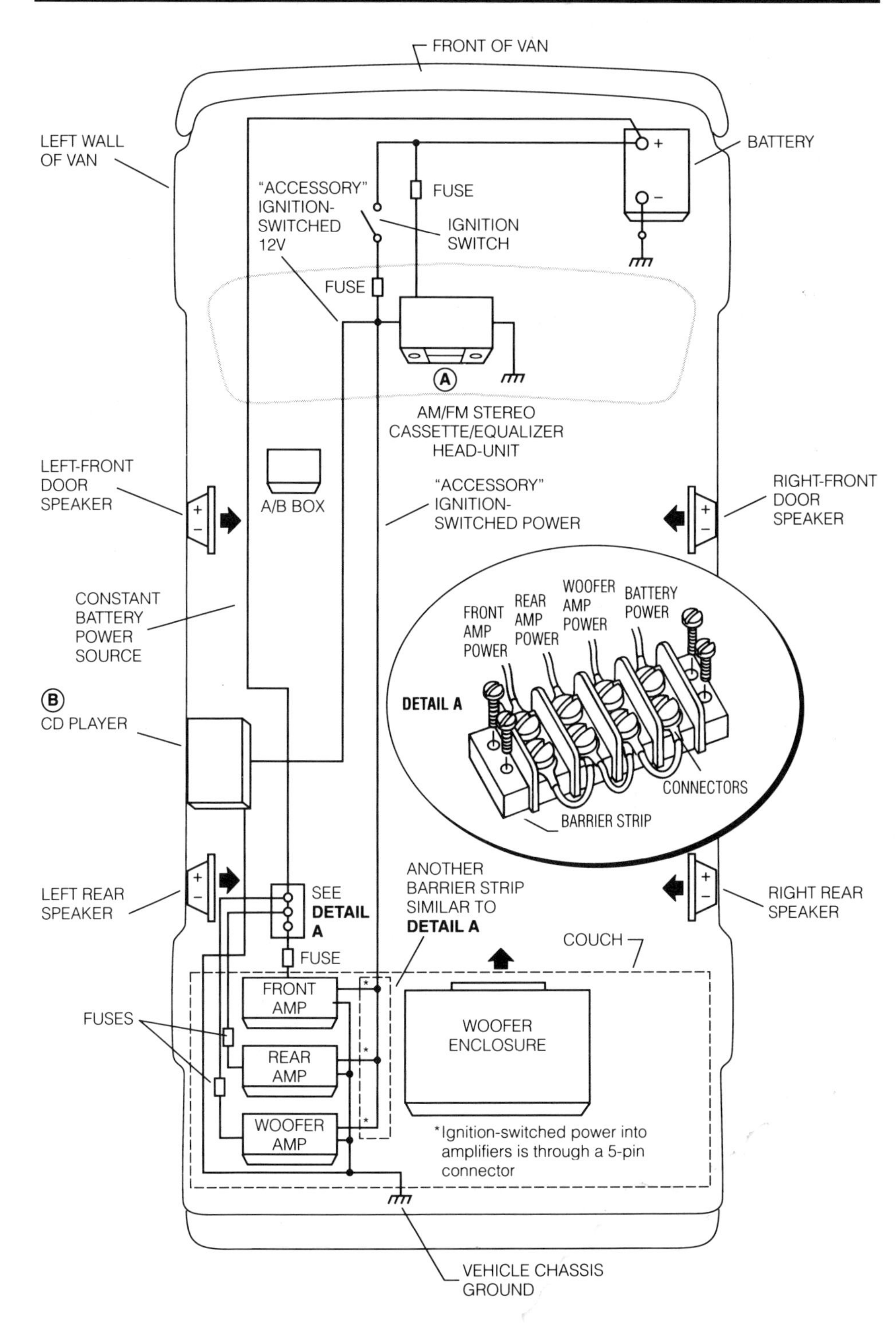

Figure 9-1. Component mounting locations and power and ground connections.

INSTALLING THE COMPONENTS

The wires were run to their appropriate destinations as the components were mounted. Some were connected immediately, but others were completed after all components were in place. Each wire that was not immediately connected was identified with a piece of tape marked with its intended connecting point.

Figure 9-2 shows a pictorial view of the system line-level and speaker output wiring to help you plan the wiring runs. The line-level inputs to the A/B control box are connected from the two sources. The outputs from the A/B control box are connected to the inputs of the three power amplifiers with phono patch cables as shown in *Figure 9-2*. The various lengths of the cables will depend on the individual installation. Audio cables in lengths of 3,' 6,' 12,' and 24' are commonly available. Measure the distances for your installation so you can select the correct length cables.

Head-Unit Installation

The DIN-C double-shaft head-unit was installed in the same way as the unit shown for a coupe/sedan in Chapter 5, *Figure 5-1.* (If you choose a DIN-E flat-face head-unit, follow the installation instructions for the unit shown in Chapter 6, *Figure 6-1.*) The van was equipped with a factory radio. After removing it from the dash and exposing the wires, we disconnected the quick-disconnects from the factory radio. The battery and ignition-switched (accessory) power leads were located as described in Chapter 3.

In the head-unit installation, we used the universal mounting kit (RS 12-1362). We carefully followed the instructions included in the mounting kit and easily adapted it to the head-unit. The installation is similar to that shown in *Figure 5-3.*

Electrical Connections

Two wires were connected to the accessory power terminal that we used for the head-unit. This is ignition-switched power. These wires are made long enough so that they extend to the rear of the van as shown in *Figure 9-1*. One of the leads will control the power amplifier switching circuit (relay) that connects the high-current constant battery power to the amplifiers. The other supplies ignition-switched power to the CD player. We made sure we had good secure ground wire connections to the van chassis for all units. Since we were not using the head-unit's speaker-level outputs, we connected the cables to the line-level outputs and ran them down behind the steering column in front of the driver compartment where we would mount the A/B control box. Finally, we bolted the head-unit in its place in the dash. Any special connectors or cables required can probably be obtained at an electronics store.

CAUTION

Always wear safety glasses when drilling, sawing and filing to prevent eye injury from flying particles.

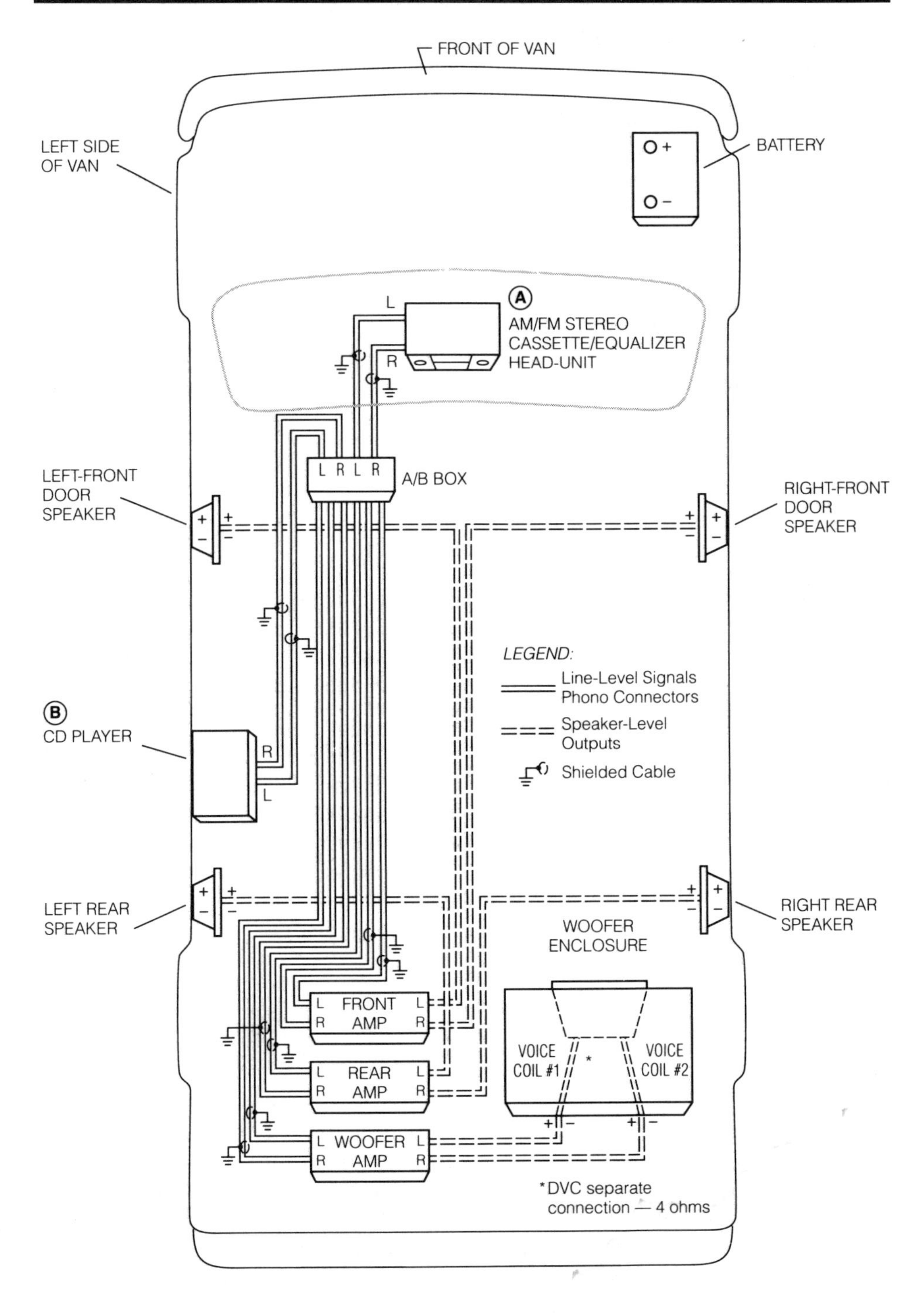

Figure 9-2. Line-level and speaker output connections.

Front Speaker Installation

Next, we mounted the surface-mount speakers on the doors facing up toward the front passengers of the van. We used the included mounting hardware to anchor them to the doors. We ran the speaker leads between the door and the main body of the van, being careful to lay them so they would not be pinched by door opening and closing. We ran the two pairs of speaker leads to the rear of the van, under the couch, where we would install the amplifiers. By using a screwdriver wrapped in cloth, you can push the speaker wires into the crevice between the carpet on the floor and the carpet on the wall. Be careful not to damage the insulation on the wires.

Rear Speaker Installation

We installed the second pair of surface-mount speakers in the rear of the van just above the rear fenders. We positioned them so the sound would project upward. We again used the mounting hardware and the same installation procedures as for the door speakers. We also ran the wires from these speakers to the amplifier location under the couch.

Amplifier Installation

CAUTION

If the amplifiers are not securely mounted, they will probably move around. Stress on the wires and components due to this motion may cause wire breakage and/or permanent damage to the amplifiers. A loose amplifier could become a dangerous missile when the vehicle comes to a sudden stop.

We installed the three amplifiers under the couch where they would be out of sight. We anchored each with self-drilling screws as shown in *Figure 5-4* and *Figure 7-3*. Since the couch was anchored to the chassis of the van, we were able to establish a good ground source by connecting the three ground leads from the amplifiers to one of the couch's anchoring bolts. For the battery source lead, we ran a 10-gauge stranded insulated wire from the battery to the amplifiers to handle the high current required by the amplifiers. We began at the amplifiers and ran it to front of the van along the left-side wall. (This is where all of the wires that go between the front and rear of the van were located.)

CAUTION

Remember to check both sides of the firewall for obstructions before drilling. Pick an existing item, such as a bolt, that goes through the firewall as a reference to determine where to locate the hole.

Once we had hidden the wire between the front and the rear of the van, we drilled a small hole slightly larger than the diameter of the wire in the firewall. We then ran the power lead through this hole and to the battery. We connected a large ring-type solderless connector to the end. We left this end disconnected from the battery until the project was completed. We sealed around the hole in the firewall with a silicone adhesive to prevent damage to the wire and to block air passage.

The battery power was distributed as shown in *Figure 9-1*, Detail A. A barrier strip with screw terminals interconnected the battery power to the three power amplifiers. It was mounted under the couch near the power amplifiers. Because line-level inputs are used for inputs to all three amplifiers, three pairs of phono cables were run between the amplifiers and the location for the A/B control box at the front of the van (see *Figure 9-2*). We left the speaker leads disconnected from the amplifiers until we constructed and installed the woofer box.

To complete the power connections, the other accessory power lead run to the rear of the van must be connected. This connection to the switching circuit of the power amplifier requires a 5-pin in- line automotive stereo connector to mate to the 5-pin connector on the power amplifier. The 5-pin connector leads carrying the power are interconnected by using another barrier strip like the one shown for the constant battery power in *Figure 9-1*, Detail A. The accessory power lead is the source and the barrier strip connections distribute the power to the amplifiers through the 5-pin connectors. When the accessory power is turned on, the amplifier is turned on. The constant battery power is connected through relay contacts that can handle the high current required by the amplifier. The ignition switch would eventually fail if it had to switch and carry this current.

CD Player Installation

We used the included "L" brackets and hardware to mount the CD player in an upright position to the wall. It was positioned for easy access by rear passengers. Its separate bass and treble controls, volume control, and CD functions permit passengers to tailor the sound in their listening area to their personal preference. The CD player's internal power amplifiers are not used. Phono cables from its line-level outputs were run to the front of the van for hookup to the A/B control box. We ran a ground lead and connected it to the same spot as the amplifiers' ground. To complete the power connection to the CD player, a male 5-pin connector mating to the 5-pin connector on the CD player is required. No signal connections are made through this 5-pin connector, just power. The power lead from the 5-pin connector, which has an in-line fuse, is connected to one of the ignition-switched power leads that we ran to the rear of the van. A quick disconnect terminal is crimped or soldered to the power lead to complete the connection. Tape the connection to insulate it.

WOOFER ENCLOSURE CONSTRUCTION

Our system includes a woofer enclosure as shown in *Figure 9-3*, which we had to construct and install. For this enclosure, we used only one 12" woofer with dual voice coil. The dual voice coils permitted us to utilize both channels of the 40W-per-channel amplifier. This allows us to pump 80 watts of power through this one speaker (it can handle 120 watts).

We designed the enclosure so it would fit under the couch in the rear of the van and be completely out of sight. The cutout patterns for the box are shown in *Figure 9-4*. Yours may require slight alteration to fit your specific application. We used 3/4" dense particle board and constructed the enclosure as described in Chapter 5. Grade A-B void-free plywood would have worked just as well. We used nettings over the back of the speaker and polyester fiber fill to dampen the enclosure.

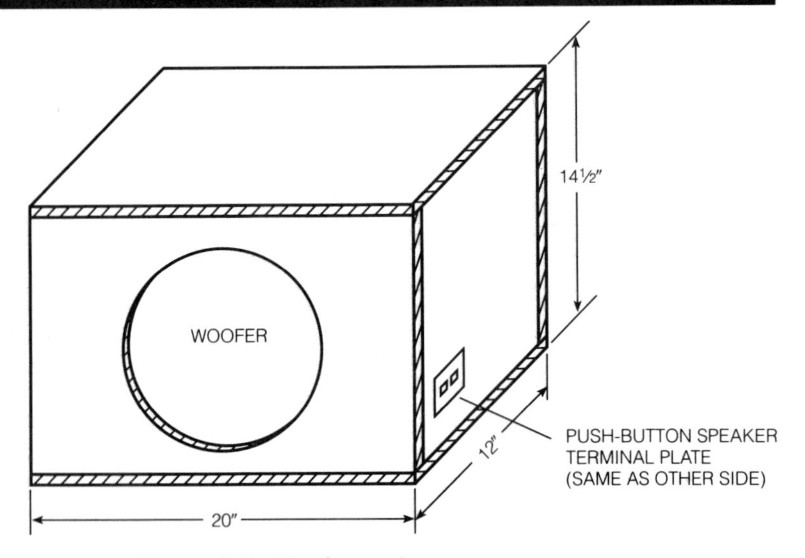

Figure 9-3. Woofer enclosure.

NOTES:
1. All material is Grade A-B "void-free" plywood or ¾″ dense particle board.
2. D is determined by selected speaker.
3. Box can be filled with synthetic fiber pillow fill to reduce standing waves and resonances in enclosure.

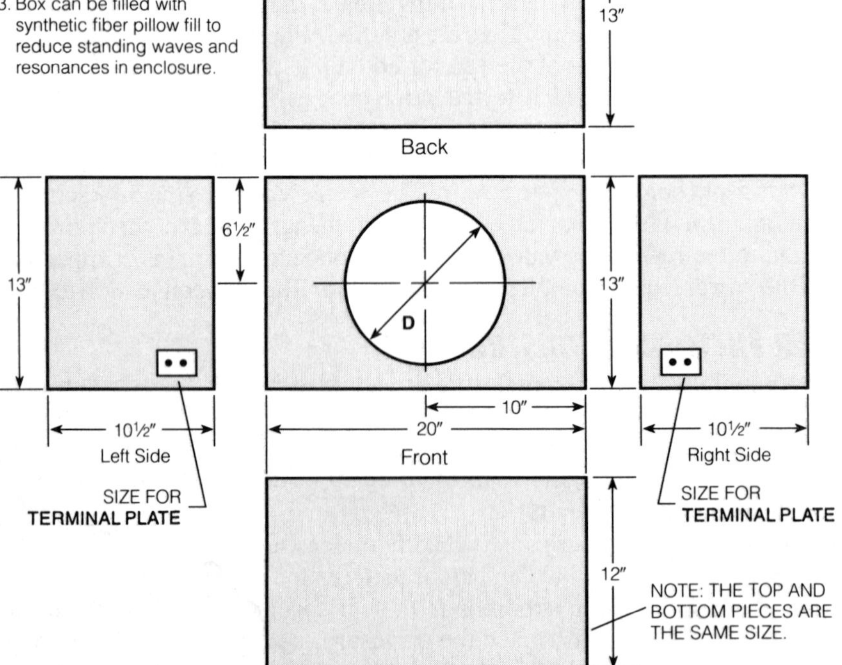

Figure 9-4. Cutting layout.

Wiring and Installation

We hooked the dual voice coils to the connection pads as shown in *Figure 9-2*. We installed the speaker and terminal plates into the box as described in Chapter 5. We did not carpet this enclosure, but painted it because it would be out of sight. We installed a wire mesh grille over the woofer to protect it since baggage, tools, and other things may be stored under the couch. We used "L" brackets to secure the box to the vehicle floor under the couch in the same way as shown in *Figure 5-7*. Being careful to observe speaker polarity, the speaker leads were connected to their appropriate amplifier outputs as shown in *Figure 9-2*.

CONSTRUCTING THE A/B CONTROL BOX

The A/B control box permits selection of the input source that is applied to each of the three amplifier/speaker systems. The three center-off DPDT switches allow any combination to be selected for the system. S1 selects the source for the woofer, S2 selects the source for the rear speakers, and S3 selects the source for the front speakers. When the switch is in the OFF position, the amplifier inputs are open. The sources in this system are a head-unit and a CD player. The system wiring is shown in *Figure 9-5,* which also shows the control box wiring. It has four input phono connectors to receive the left and right outputs of the two sources, and six output phono connectors to feed the source signals to the inputs of the left and right channels of the three amplifiers.

You will need the following tools to build this switch box:

- Hand drill and drill bits
- Soldering iron and solder
- Knife
- Long-nose pliers
- Wire cutters

The parts needed for construction of the control box are as follows:

Qty	Item
1	Experimenter Enclosure
3	Flat-lever toggle switches DPDT C-off
1	Eight-position phono jack board
1	Four-position phono jack board
8	Screws 4-40, 1/4″ long
8	Nuts 4-40
	Hook-up wire, #20 gauge
	Solid bus wire, #20 gauge

The first step is to drill the holes in the box for mounting the switches and the phono jacks, and the two holes used to fasten the A/B control box to the bottom of the van's dash. To determine where to position the dash mounting holes, you may need to experiment by holding the box at various locations to find an appropriate mounting space in the van. We mounted the box under the dash on the left side, so it would be accessible to the driver.

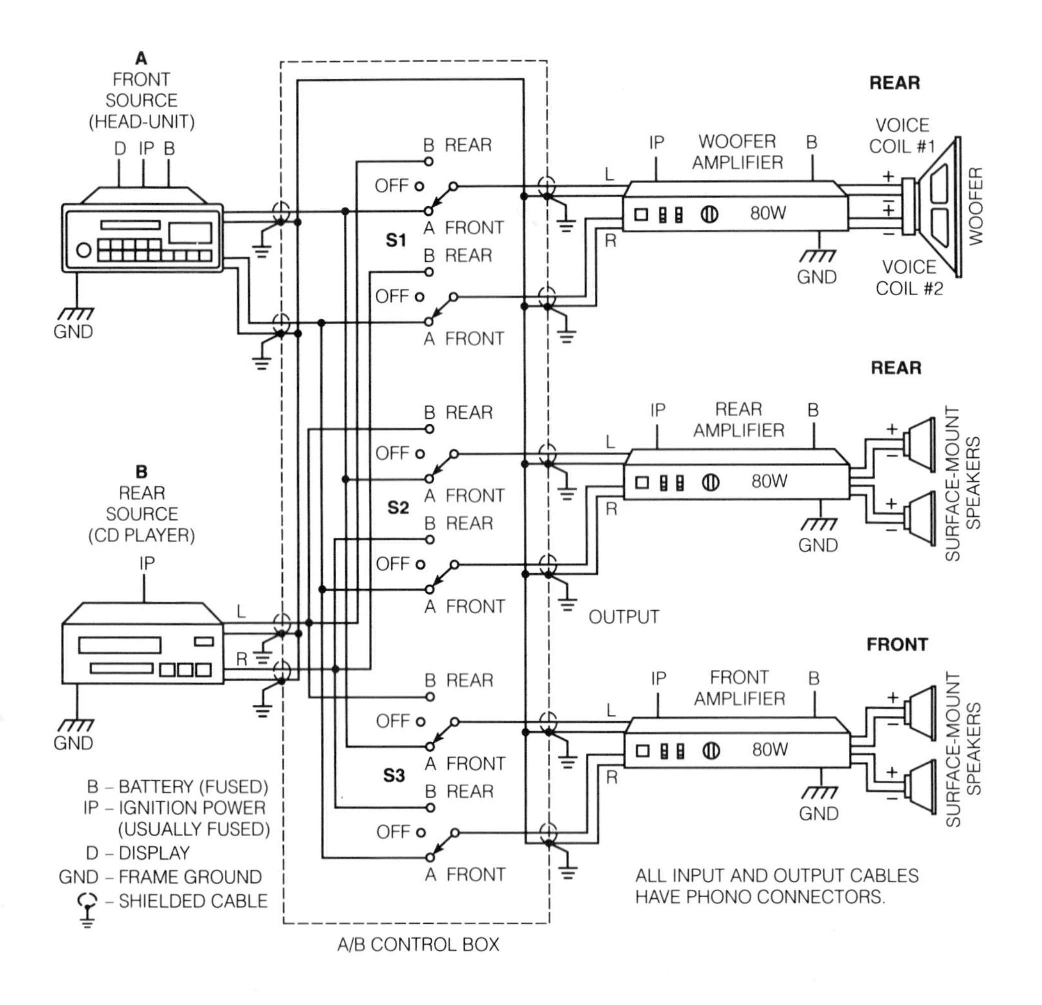

Figure 9-5. System wiring diagram for 1988 Dodge Van.

CAUTION

Be careful when you are locating the place to drill the 9/32" mounting holes so that you do not damage something with the drill.

Drill the holes according to the layouts shown in *Figure 9-6a, 9-6b,* and *9-6c.* Mount the eight-position and the four-position phono jack boards to the box using 4-40, 1/4" screws and nuts. The four-position jacks will be the control box inputs from the front and rear signal sources and the eight-position jack board will be the control box outputs to the amplifiers (we will use only six of the eight jacks). Mount the three switches and the two phono jack boards in the control box.

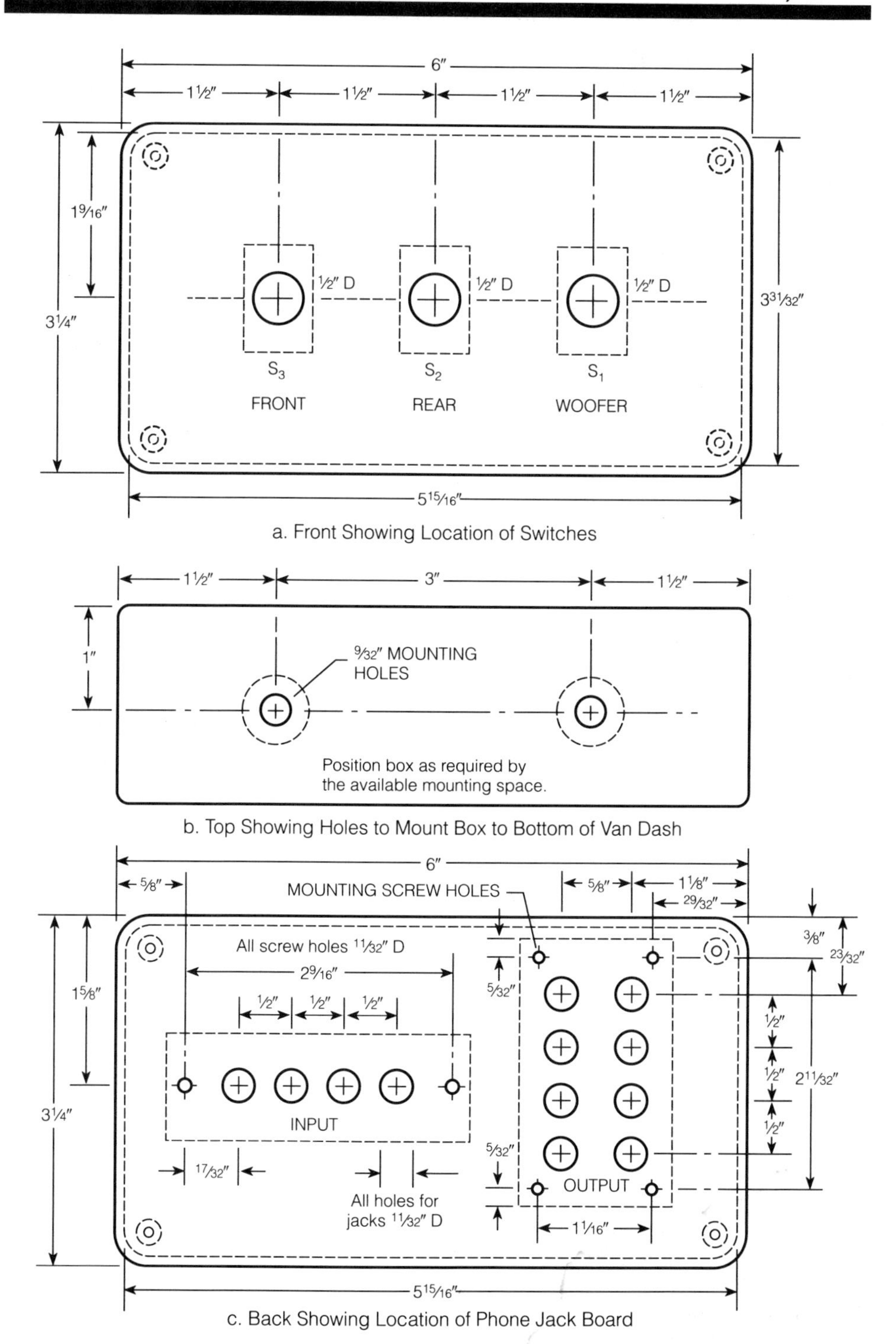

a. Front Showing Location of Switches

b. Top Showing Holes to Mount Box to Bottom of Van Dash

c. Back Showing Location of Phone Jack Board

Figure 9-6. Hole locations for A/B control box.

Wiring and Installation

The ground terminals on the phono jacks will be the tie-points for the common connection of all the grounds. Insert a length of solid bus wire so that it goes through each of the ground terminals of both the input and the output jacks as shown in *Figure 9-7a.* Carefully solder each one of the ground terminals. Poor solder joints here can cause a very annoying noise problem in the sound system.

Connect the signal leads of the output and input jacks to the toggle switches as shown in *Figures 9-7b* and *9-7c.* Solder these connections carefully. The A/B control box is now ready to mount in the van and be connected into the system.

Mount the box with 10-32 bolts with flat washers, lock washers and nuts. Attach the back with the self-tapping screws provided and connect the phono connectors of the cables that were run to the respective jacks at the back of the A/B box as shown in *Figure 9-2.* Run the remaining speaker wires if any still need to be connected. Check all connections again according to *Figure 9-2* and *9-5.* Connect the amplifiers' constant power wire and the vehicle battery cable to the battery and the system is ready to use.

This system provides great versatility, but as we have emphasized throughout this book, you can personalize it to suit your specific desires.

SUMMARY

In this book, we have provided descriptions of actual automobile sound installations in particular vehicles. We have tried to make the installation instructions sufficiently broad so that they can be applied to a wide variety of similar types of vehicles. We have tried to include enough detail to get you to understand what components are required, how they are interconnected, and how to get started. We hope that we have succeeded.

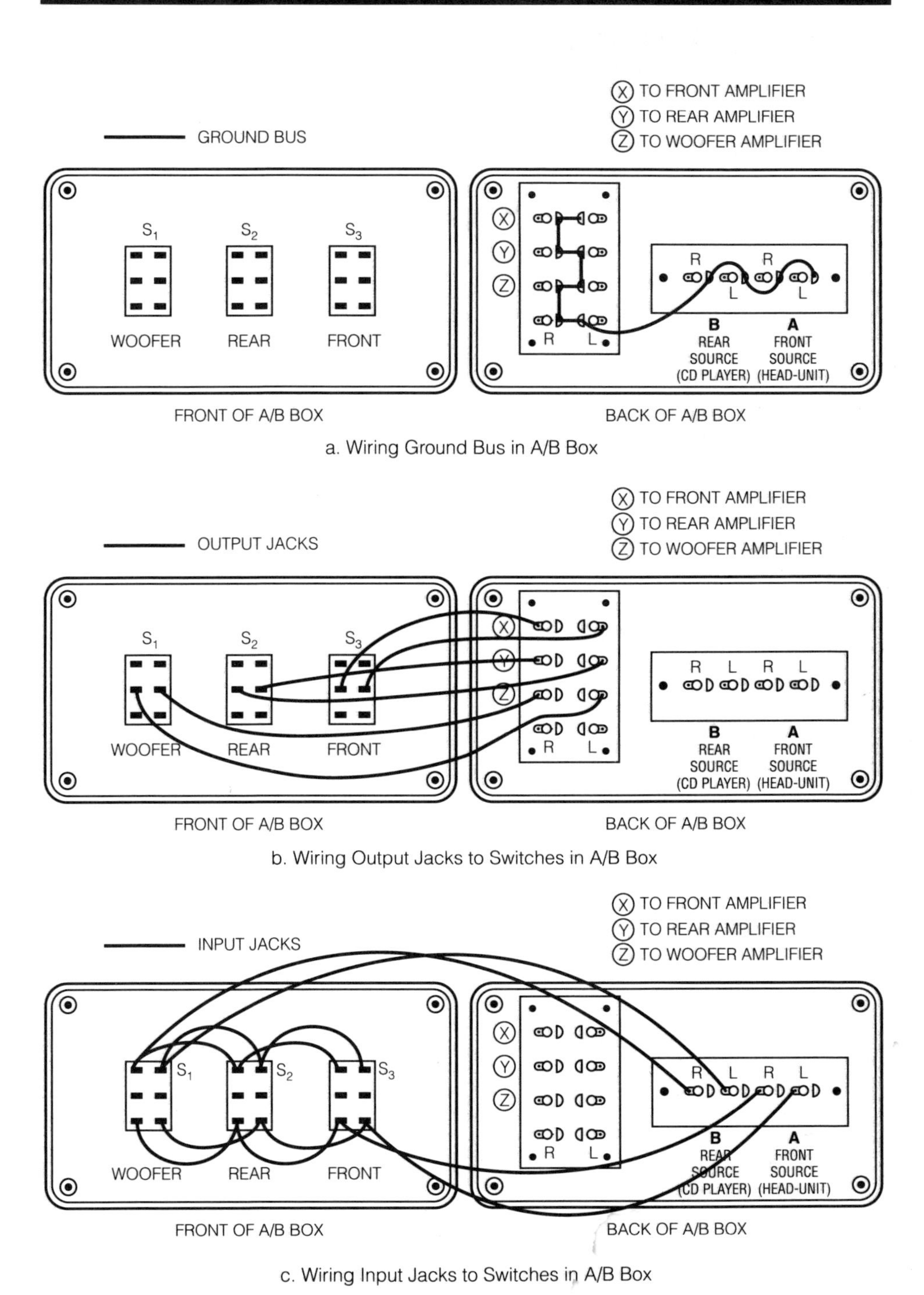

Figure 9-7. Wiring in A/B control box.

APPENDIX

A. Logarithms

EXPONENTS

A logarithm (log) is the exponent (or power) to which a given number, called the base, must be raised to equal the quantity. For example:

Since $10^2 = 100$, then the log of 100 to the base 10 is equal to 2, or $\text{Log}_{10}\ 100 = 2$

Since $10^3 = 1000$, then the log of 1000 to the base 10 is equal to 3, or $\text{Log}_{10}\ 1000 = 3$

BASES

There are three popular bases in use—10, 2 and ϵ. Logarithms to the base 10 are called common logarithms (log). Logarithms in base ϵ are called natural logarithms (ln).

Logarithms to the base 2 are used extensively in digital electronics.

Logarithms to the base ϵ (approximately 2.71828...) are quite frequently used in mathematics, science and technology. Here are examples:

Base 10

$\log_{10}2 = 0.301$ is $10^{0.301} = 2$

$\log_{10}200 = 2.301$ is $10^{2.301} = 200$

Base 2

$\log_2 8 = 3$ is $2^3 = 8$

$\log_2 256 = 8$ is $2^8 = 256$

Base ϵ

$\text{in}_\epsilon 2.71828 = 1$ is $\epsilon^1 = 2.71828$

$\text{in}_\epsilon 7.38905 = 2$ is $\epsilon^2 = 7.38905$

RULES OF EXPONENTS

Since a logarithm is an exponent, the rules of exponents apply to logarithms:

$$\log (M \times N) = (\log M) + (\log N)$$

$$\log (M/N) = (\log M) - (\log N)$$

$$\log M^N = N \log M$$

B. Decibels

The bel is a logarithmic unit used to indicate a ratio of two power levels (sound, noise or signal voltage, microwaves). It is named in honor of Alexander Graham Bell (1847-1922) whose research accomplishments in sound were monumental. A 1 bel change in strength represents a change of ten times the power ratio. In normal practice, the bel is a rather large unit, so the decibel (dB), which is 1/10 of a bel, is commonly used.

$$\text{Number of dB} = 10 \log P2/P1$$

A 1 dB increase is an increase of 1.258 times the power ratio, or 1 db = 10 log 1.258.

A 10 dB increase is an increase of 10 times the power ratio, or 10 db = 10 log 10.

Other examples are:

3 dB = 2 times the power ratio

20 dB = 100 times the power ratio

−30 dB = 0.001 times the power ratio

It is essential to remember that the decibel is *not* an absolute quantity. It merely represents a change in power level relative to the level at some different time or place. It is meaningless to say that a given amplifier has an output of so many dB unless that output is referred to a specific power level. If we know the value of the input power, then the *ratio* of the output power to the specific input power (called power gain) may be expressed in dB.

If a standard reference level is used, then *absolute power* may be expressed in dB *relative* to that standard reference. The commonly used reference level is one milliwatt. Power referenced to this level is expressed in dBm. Here are power ratios and dBm ratios:

dB	Power Ratio	dBm	Power (mw)
1	1.258	1	1.258
3	2	3	2
10	10	10	10
20	100	20	100
–30	0.001	–30	0.001

C. Metric Conversions

International System of Units (SI) — Metric Units

Prefix	Symbol		Multiplication Factor
exa	E	10^{18} =	1,000,000,000,000,000,000
peta	P	10^{15} =	1,000,000,000,000,000
tera	T	10^{12} =	1,000,000,000,000
giga	G	10^{9} =	1,000,000,000
mega	M	10^{6} =	1,000,000
kilo	k	10^{3} =	1,000
hecto	h	10^{2} =	100
deca	da	10^{1} =	10
(unit)		10^{0} =	1
deci	d	10^{-1} =	0.1
centi	c	10^{-2} =	0.01
milli	m	10^{-3} =	0.001
micro	u	10^{-6} =	0.000001
nano	n	10^{-9} =	0.000000001
pico	p	10^{-12} =	0.000000000001
femto	f	10^{-15} =	0.000000000000001
atto	a	10^{-18} =	0.000000000000000001

1 meter (m) =
100 centimeters (cm) =
1000 millimeters (mm)

25.4	mm	=	1 inch
2.54	cm	=	1 inch
30.48	cm	=	1 foot
0.3048	m	=	1 foot
0.9144	m	=	1 yard
1.609	km	=	1 mile
1.852	km	=	1 nautical mile

Fractional Dimensions

Inches		Millimeters	Inches		Millimeters	Inches		Millimeters
1/64	0.016	0.397	23/64	0.359	9.128	11/16	0.688	17.463
1/32	0.031	0.794	3/8	0.375	9.525	45/64	0.703	17.859
3/64	0.047	1.191	25/64	0.391	9.922	23/32	0.719	18.256
1/16	0.063	1.588	13/32	0.406	10.319	47/64	0.734	18.653
5/64	0.078	1.984	27/64	0.422	10.716	3/4	0.750	19.050
3/32	0.094	2.381	7/16	0.438	11.113	49/64	0.766	19.447
7/64	0.109	2.778	29/64	0.453	11.509	25/32	0.781	19.844
1/8	0.125	3.175	15/32	0.469	11.906	51/64	0.797	20.241
9/64	0.141	3.572	31/64	0.484	12.303	13/16	0.813	20.638
5/32	0.156	3.969	1/2	0.500	12.700	53/64	0.828	21.034
3/16	0.188	4.762	33/64	0.516	13.097	27/32	0.844	21.431
13/64	0.203	5.159	17/32	0.531	13.494	55/64	0.859	21.828
7/32	0.219	5.556	35/64	0.547	13.891	7/8	0.875	22.225
15/64	0.234	5.953	9/16	0.563	14.288	57/64	0.891	22.622
1/4	0.250	6.350	37/64	0.578	14.684	29/32	0.906	23.019
17/64	0.266	6.747	19/32	0.594	15.081	59/64	0.922	23.416
9/32	0.281	7.144	39/64	0.609	15.478	15/16	0.938	23.813
19/64	0.297	7.541	5/8	0.625	15.875	61/64	0.953	24.209
5/16	0.313	7.938	41/64	0.641	16.272	31/32	0.969	24.606
21/64	0.328	8.334	21/32	0.656	16.669	1.0	1.000	25.400
11/32	0.344	8.731	43/64	0.672	17.066			

GLOSSARY

acoustic feedback: A squealing sound when the output of an audio circuit is fed back in phase into the circuit's input.
acoustic fiberfill: Thin fiberglass or polyester fiberfill material used as damping material inside speaker enclosures. Typically available at sewing centers for stuffing pillows.
acoustic suspension: A speaker designed for, or used in, a sealed enclosure.
ac coupling: Coupling between electronic circuits that passes only alternating current and time varying signals, not direct current.
acoustics: The science or study of sound.
air suspension: An acoustic suspension speaker.
alternating current (ac): An electrical current that periodically changes in magnitude and direction.
ambience: A surround or concert-hall sound.
ampere (A): The unit of measurement for electrical current in coulombs (6.25×10^{18} electrons) per second. There is one ampere in a circuit that has one ohm resistance when one volt is applied to the circuit. See Ohm's law.
amplifier: An electrical circuit designed to increase the current, voltage, or power of an applied signal.
amplitude: The relative strength (magnitude) of an electrical signal above or below some reference, often ground or zero. Amplitude can be expressed as either a negative or positive number, depending on the signals being compared; usually measured in volts or amperes.
attenuation: The reduction, typically by some controlled amount, of an electrical signal.
audio frequency: The acoustic spectrum of human hearing, generally regarded to be between 20 Hz and 20,000 Hz.
baffle: A piece of wood inside an enclosure used to direct or block the movement of sound.
balance: Equal signal strength provided to both left and right stereo channels.
bandpass filter: An electric circuit designed to pass only middle frequencies. See also high-pass filter and low-pass filter.
basket: The metal frame of a speaker.
bass: The low end of the audio frequency spectrum: approximately 20 Hz to about 1000 Hz.
bass reflex: A ported reflex speaker enclosure.
battens: Small strips of wood placed inside a speaker to reinforce its mating corners or to provide a mounting surface for front and back panels.
bobbin: A paper, plastic, or metal cylinder around which is wound the wire that forms a speaker's voice coil. The bobbin is mechanically connected to the speaker cone.
capacitor (C): A device made up of two metallic plates separated by a dielectric (insulating material). Used to store electrical energy in the electrostatic field between the plates. It produces an impedance to an ac current.
cassette: 1. The two-reel plastic carrier that contains audio magnetic tape. 2. The shortened name for the automotive sound system component that plays the audio magnetic tape contained in cassettes.
CD: Compact disc or the compact disc player.
channel: The left or right signals of a stereo audio system.
circuit: A complete path that allows electrical current from one terminal of a voltage source to the other terminal.

clipping: A distortion caused by cutting off the peaks of audio signals. Clipping usually occurs in the amplifier when its input signal is too high or when the volume control is turned up too high.

coaxial driver: A speaker that is composed of two individual voice coils and cones; used for reproduction of sounds in two segments of the sound spectrum. See also triaxial driver.

coloration: "Smearing" sounds by adding frequencies due to intermodulation distortion. More prevalent at high audio frequencies.

compliance: The relative stiffness of a speaker suspension, typically indicated simply as "high" or "low," but technically specified as Vas.

cone: The cone-shaped diaphragm of a speaker attached to the voice coil. It produces pulsations of air that the ear detects as sound.

crossover network: An electric circuit or network that splits the audio frequencies into different bands for application to individual speakers.

current (I): The flow of charge measured in amperes.

damping: 1. Acoustic fiberglass or polyester fiberfill material used inside speaker enclosures. 2. The reduction of movement of a speaker cone, due either to the electromechanical characteristics of the speaker driver and suspension, or the effects of pressure inside a speaker enclosure.

decibel (dB): A logarithmic scale used to denote a change in the relative strength of an electric signal or acoustic wave. It is a standard unit for expressing the ratio between power level P_1 and power level P_2. $dB = 10 \log_{10} P_1/P_2$. An increase of 3 dB is a doubling of electrical (or signal) power; an increase of 10 dB is a doubling of perceived loudness. The decibel is not an absolute measurement, actually, but indicates the relationship or ratio between two signal levels. See Appendix.

DIN-C: A set of standard specifications for automotive head-units with two shaft openings and a rectangular opening.

DIN-E: A set of standard specifications for automotive head-units with a single rectangular opening. Commonly called "flat-face" design.

direct current (dc): Current in only one direction.

dispersion: The spreading of sound waves as they leave a speaker.

distortion: Any undesirable change in the characteristics of a reproduced audio signal that degrades the fidelity of the original signal.

dolby: A type of dynamic noise reduction system.

dome tweeter: A high frequency speaker with a dome-shaped diaphragm that provides much better dispersion of high frequencies than standard cone speakers.

driver: Technically, the electromagnetic components of a speaker, typically consisting of a magnet and voice coil, but also used to describe any direct radiator speaker.

dual voice-coil: Two voice coils wound on the same bobbin and driving the same cone.

ducted port: A ported reflex speaker enclosure.

dynamic-range: The range of sound levels, from softest to loudest, which a system can reproduce without distortion. Usually expressed in decibels.

enclosure: A housing for any electrical or electronic device; specifically used in this book for a box enclosing speakers.

equalizer: An adjustable audio filter inserted in a circuit or system to divide and adjust its frequency response.

equalization: As used in audio, the adjustment of frequency response to tailor the sound to match personal preferences, room acoustics, and speaker enclosure design.

fader: A variable control used to change the distribution of power between front and rear speakers.

farad: The basic unit of capacitance. A capacitor has a value of one farad when it can store one coulomb of charge with one volt across it.

fidelity: A measure of how true a circuit, amplifier, system or subsystem reproduces its input signal.

filter: An electrical circuit designed to prevent or reduce the passage of certain frequencies.

flat response: The faithful reproduction of an audio signal; specifically, variations in output level of less than one decibel above or below a median level over the audio spectrum.

free-air resonance: The natural resonant frequency of a woofer speaker when operating outside an enclosure.

frequency: The number of waves (or cycles) arriving at or passing a point in one second; expressed in hertz (or Hz).

frequency response: A plot of how a range of frequencies are faithfully reproduced by a given speaker or audio system.

fundamental or fundamental tone: The tone produced by the lowest frequency component of an audio signal.

full-range: A speaker designed to reproduce all or most of the sound spectrum.

golden ratio: The ratio of the depth, width, and height of a speaker enclosure, based on the Greek Golden Rectangle, and which most often provides the best sound. W = 1.0, Depth = 0.6W, Height = 1.6W

grille cloth: Fabric used to cover the speaker mounted in an enclosure.

ground: Refers to a point of (usually) zero voltage, and can pertain to a power circuit or a signal circuit. Commonly used as the reference point for voltage measurements.

harmonic: The multiple frequencies of a given sound, created by the interaction of signal waveforms. A "middle C" on the piano has a fundamental audio frequency of 256 Hz, but also a number of secondary higher frequencies (harmonics) that are odd and even multiples of this fundamental.

harmonic distortion: Harmonics artificially added by an electrical circuit or speaker, and are generally undesirable. It is expressed as a percentage of the original signal.

head-unit: The name given to automotive sound system component that supplies the main signal source from AM, FM, cassette tape or CD player.

hertz: A unit of frequency equal to one cycle per second, named after German physicist H.R. Hertz.

high-fidelity: Commonly called hi-fi, it refers to the reproduction of sound with little or no distortion.

high-pass filter: An electric circuit designed to pass only high frequencies. See also bandpass filter and low-pass filter.

hiss: Audio noise that sounds like air escaping from a tire.

horn: A speaker design using its own funnel-shaped conduit to amplify, disperse, or modify the sounds generated by the internal diaphragm of the speaker.

hum: Audio noise that has a steady low frequency pitch, typically caused by the effects of induction by nearby ac lines or leakage of ac line frequency into an amplifier's signal circuits.

impedance: The opposition of a circuit or speaker to an alternating current.

inductance (L): The capability of a coil to store energy in a magnetic field surrounding it. It produces an impedance to an ac current.

intensity: The strength of a sound signal represented by the amplitude of the pressure waves producing the sound. Expressed in watts per square meter.

intermodulation distortion: Combinations of two or more frequencies which generate new frequencies which are sums and differences of the original signal.

L-pad: A type of potentiometer that maintains constant impedance at its input while varying the signal level at its output. L-pads are most often used as an external balance control or variable attenuator (volume control).

loudness: A measure of the sensitivity of human hearing to the strength of sound.

low-pass filter: An electric circuit designed to pass only low frequencies. See also bandpass filter and high-pass filter.

microphone: A device that converts sound waves into electrical signals.

midrange: A speaker designed to reproduce the middle frequencies of the sound spectrum, generally most efficient between about 1000 Hz to 4000 Hz.

mounting flange: The outer edges of a speaker frame which has pre-drilled holes to accept screws or bolts for securing it to the enclosure.

noise: Any unwanted signal or electromagnetic radiation, particularly that which distorts signals or disrupts normal operation.

noise factor/figure: For a given bandwidth, the ratio of total noise power ratio at the input to the noise power ratio at the output when the ratio is expressed in dB, it is noise figure.

octave: The logarithmic relation of sound frequencies used by musicians. The frequency of each higher octave is twice the preceding one.

ohm (W): A unit of electrical resistance or impedance.

Ohm's law: A basic law of electric circuits. It states that the current I in amperes in a circuit is equal to the voltage E in volts divided by the resistance R in Ohms; thus, I = E/R.

passive radiator (or drone): A speaker with a cone but no driver components. The cone vibrates with the change in pressure inside the speaker enclosure. Typically used to increase bass output with no increase in electrical power.

peak: The maximum amplitude of a voltage or current.

period: For electronic circuits, the length of time required for one cycle of a periodic wave.

phase: The angular or time displacement between the voltage and current in an ac circuit.

phase distortion: The distortion which occurs when one frequency component of a complex input signal takes longer to pass through an amplifier or system than another frequency.

pitch: How a tone sounds to the human ear. It is a subjective term because the perception of frequency varies with sound intensity.

piezoelectric: A characteristic of some materials, especially crystal, that when subjected to electric voltage the material vibrates. Sometimes used in tweeters in place of a magnet, voice coil, and cone.

polarity: The orientation of magnetic or electric fields. The polarity of the incoming audio signal determines the direction of movement of the speaker cone.

ported reflex: A type of speaker enclosure that uses a duct or port to improve efficiency at low frequencies.

power: The time rate of doing work or the rate at which energy is used. A watt of electrical power is the use of one joule of energy per second. Watts of electrical energy equals volts times amperes.

resonance: The tendency of a speaker to vibrate most at a particular frequency; sometimes referred to as natural frequency.

resistance: In electrical or electronic circuits, a characteristic of a material that opposes the flow of electrons. It results in loss of energy in a circuit dissipated as heat. Speakers have resistance that opposes current.

RMS: An acronym for root mean square. The RMS value of an alternating current produces the same heating effect in a circuit as the same value of a direct current.

selectivity: The characteristic that describes the ability of a tuned circuit or a receiver to select the signal frequencies desired and reject the ones not desired.

sensitivity: A measure of the ability of a receiver to amplify weak signals.

signal: The desired portion of electrical information.

signal-to-noise (S/N): The ratio, expressed in dB, between the signal (sound you want) and noise (sound you don't want).

sine wave: The waveform of a pure alternating current or voltage. It deviates about a zero point to a positive value and a negative value. Audio signals are sine waves or combinations of sine waves.

sound: The vibratory energy of air particles. The signals of frequencies from 20 to 20 kHz that normally are detected by the human ear.

sound pressure level (SPL): The loudness of an acoustic wave stated in dB that is proportional to the logarithm of its intensity.

spectrum: The complete range of frequencies from the lowest to highest.

spider: The flexible fabric that supports the bobbin, voice coil, and inside portion of the cone within the speaker frame.

static: Random noise in a communications system receiver due to atmospheric or manmade electrical disturbances.

subwoofer: A speaker especially designed to handle the bass portion (20 Hz to 150 Hz) of the audio spectrum.

surround: The outer suspension of a speaker cone; the surround connects the outside portion of the cone to the speaker frame.

suspension: See surround.

three-way: A type of speaker system composed of three ranges of speakers, specifically a tweeter, midrange, and woofer. See also two-way.

timbre: A subjective term used for human hearing that gives a sound a particular identity. It is related to the spectrum of frequencies contained within the sound.

total harmonic distortion (THD): The percentage, in relation to a pure input signal, of harmonically derived frequencies introduced in the sound reproducing circuitry and hi-fi equipment (including the speakers).

transient response: The instantaneous change is an electronic circuit's output response when input circuit conditions suddenly change from one steady-state condition to another.

transient intermodulation distortion: A distortion which occurs principally during loud, high-frequency music passages in solid-state amplifiers that use large amounts of negative feedback.

treble: The upper end of the audio spectrum, usually reproduced by a tweeter.

triaxial driver: A speaker that is composed of three individual voice coils and cones; used for the reproduction of sounds in three segments of the sound spectrum. See also coaxial driver.

tuner: The part of a radio or TV receiver containing the rf amplifier, mixer and local oscillator that selects the desired signal.

tweeter: A speaker designed to reproduce the high or treble range of the sound spectrum, generally most efficient from about 4000 Hz to 20,000 Hz.

two-way: A type of speaker system composed of two ranges of speakers, consisting of any two of the following: a tweeter, midrange, and woofer. See also three-way. Some midrange speakers are classified as midrange/tweeter.

voice coil: The wire wound around the speaker bobbin. The bobbin is mechanically connected to the speaker cone and causes the cone to vibrate in response to the audio current in the voice coil.

watt: A unit of electrical power.

wavelength: The distance a wave travels in the time required to complete one cycle.

whizzer: A small supplementary cone attached to the center of the speaker's main cone for the purpose of increasing high frequency response.

woofer: A speaker designed to reproduce the low frequencies of the sound spectrum, generally most efficient from about 20 Hz to 1000 Hz.

INDEX

Dear Reader: ***We'd like your views on the books we publish.***

PROMPT® Publications, a division of Howard W. Sams & Company (A Bell Atlantic Company), is dedicated to bringing you timely and authoritative documentation and information you can use. You can help us in our continuing effort to meet your information needs. Please take a few moments to answer the questions below. Your answers will help us serve you better in the future.

1. What is the title of the book you purchased?____________________
2. Where do you usually buy books?____________________
3. Where did you buy this book?____________________
4. What did you like most about the book?____________________
5. What did you like least?____________________
6. Is there any other information you'd like included?____________________
7. In what subject areas would you like us to publish more books? (Please check the boxes next to your fields of interest.)

- ☐ Audio Equipment Repair
- ☐ Camcorder Repair
- ☐ Computer Repair
- ☐ Electronic Concepts Theory
- ☐ Electronic Projects/Hobbies
- ☐ Electronic Reference
- ☐ Home Appliance Repair
- ☐ Mobile Communications
- ☐ Security Systems
- ☐ Sound System Installation
- ☐ TV Repair
- ☐ VCR Repair

8. Are there other subjects that you'd like to see books about? ____________________

9. Comments ____________________

Name ____________________

Address ____________________

City ____________________ State/ZIP __________

E-Mail ____________________

Would you like a ***FREE*** PROMPT® Publications catalog? ☐ Yes ☐ No

Thank you for helping us make our books better for all of our readers. Please drop this postage-paid card into the nearest mailbox.

For more information about PROMPT® Publications, see your authorized Howard Sams distributor or call 1-800-428-7267 for the name of your nearest PROMPT® Publications distributor.

PROMPT® PUBLICATIONS

A Division of ***Howard W. Sams & Company***
A Bell Atlantic Company
2647 Waterfront Parkway, East Dr.
Indianapolis, IN 46214-2041